TECHNOLOGY, TRADE, AND THE U.S. ECONOMY

Report of a Workshop Held at
Woods Hole, Massachusetts, August 22-31, 1976

Conducted by the Office of the Foreign Secretary,
National Academy of Engineering, and
Assembly of Engineering,
National Research Council

NATIONAL ACADEMY OF SCIENCES
Washington, D.C. 1978

NOTICE: The project that is the subject of this report was approved by the Governing Board of the National Research Council, whose members are drawn from the Councils of the National Academy of Sciences, the National Academy of Engineering, and the Institute of Medicine. The members of the Committee responsible for the report were chosen for their special competences and with regard for appropriate balance.

This report has been reviewed by a group other than the authors according to procedures approved by a Report Review Committee consisting of members of the National Academy of Sciences, the National Academy of Engineering, and the Institute of Medicine.

This project was supported under Contract No. NSF-C310, T.O. 324, between the National Science Foundation and the National Academy of Sciences.

Library of Congress Catalog Card Number 78-59758

International Standard Book Number 0-309-02761-6

Available from

Printing and Publishing Office
National Academy of Sciences
2101 Constitution Avenue, N.W.
Washington, D.C. 20418

Printed in the United States of America

PARTICIPANTS AT A WORKSHOP ON TECHNOLOGICAL FACTORS CONTRIBUTING TO THE NATION'S FOREIGN TRADE POSITION, Woods Hole, Massachusetts, August 22-31, 1976

Committee

MILTON KATZ, *Workshop Chairman*, Henry L. Stimson Professor of Law and Director, International Legal Studies, Harvard Law School
N. BRUCE HANNAY, *Study Chairman*, Foreign Secretary, National Academy of Engineering, and Vice-President, Research and Patents, Bell Laboratories
JACK BARANSON, President, Developing World Industry and Technology, Inc.
LEWIS M. BRANSCOMB, Vice-President and Chief Scientist, IBM Corporation
GEORGE COLLINS, Assistant to the President and Director of International Affairs, International Union of Electrical, Radio, and Machine Workers
JOHN T. DUNLOP, School of Business Administration, Harvard University
PETER T. JONES, Vice-President, Legal and Government Affairs and General Counsel, Legal Division, Marcor, Inc.
CHARLES P. KINDLEBERGER, Ford Professor of Economics, Massachusetts Institute of Technology
RALPH LANDAU, Chairman of the Board and Chief Executive Officer, Halcon International, Inc.
C. W. MAYNES, Secretary, Carnegie Endowment for International Peace
NATHANIEL SAMUELS, Chairman of the Board, Louis Dreyfus Holding Company, Inc., and Chairman of the Board of Advisory Directors, Kuhn, Loeb, and Co., Inc.
BENJAMIN A. SHARMAN, International Trade Representative, International Association of Machinists and Aerospace Workers
LOWELL W. STEELE, Consultant--Planning, General Electric Company

RAYMOND VERNON, Director, Center for International Affairs, Harvard University

Additional Participants

BETSY ANCKER-JOHNSON, Assistant Secretary of Commerce for Science and Technology, U.S. Department of Commerce
DAVID BECKLER, Assistant to the President, National Academy of Sciences
MICHAEL BORETSKY, Senior Policy Analyst, Office of the Secretary, U.S. Department of Commerce
J. FRED BUCY, President, Chief Operating Officer, and Director, Texas Instruments
JOHN GRANGER, Deputy Assistant Director for Scientific, Technological, and International Affairs, National Science Foundation
EDGAR HARRELL, Director, Planning and Analysis Staff, Bureau of Economic and Business Affairs, U.S. Department of State
GEORGE R. HEATON, Research Associate, Center for Policy Alternatives, Massachusetts Institute of Technology
J. HERBERT HOLLOMON, Director, Center for Policy Alternatives, Massachusetts Institute of Technology
GARY C. HUFBAUER, International Tax Staff, U.S. Treasury
STANLEY KATZ, Deputy Assistant Secretary for International Economic and Policy Research, U.S. Department of Commerce
REGINA KELLY, Special Assistant to the Director, Office of Economic Research, U.S. Department of Commerce
HERBERT S. LEVINE, Professor of Economics, Faculty of Arts and Science, University of Pennsylvania, and Stanford Research Institute
HYLAN LYON, Deputy Director, Science, Technology, and Industry, Organization for Economic Cooperation and Development
HENRY R. NAU, Special Assistant, Office of the Undersecretary for Economic Affairs, U.S. Department of State, and Professor of Political Science, The George Washington University
SUMIYE OKUBO, Policy Analyst, Division of Policy Research and Analysis, National Science Foundation
ROLF PIEKARZ, Senior Policy Analyst, Division of Policy Research and Analysis, National Science Foundation
HERMAN POLLACK, Office of Technology Assessment, Congress of the United States, and Research Professor of International Affairs, The George Washington University

ROGER SHIELDS, Deputy Assistant Secretary of Defense, U.S. Department of Defense
HELENA STALSON, Staff Economist, Council on Foreign Relations
GUS WEISS, Council on International Economic Policy, The White House

Staff

HUGH H. MILLER, *Study Director* and Executive Director, Office of the Foreign Secretary, National Academy of Engineering
E. M. (MONTY) GRAHAM, *Associate Study Director and Author of Report for Committee* and Assistant Professor of Management, Alfred P. Sloan School of Management, Massachusetts Institute of Technology
JULES SCHWARTZ, *Rapporteur* and Associate Professor of Management, The Wharton School, University of Pennsylvania
WILLIAM DAVIDSON, *Assistant Rapporteur* and Research Associate, Graduate School of Business Administration, Harvard University
KERSTIN B. POLLACK, *Study Assistant* and Associate Director, New Programs Development, Assembly of Engineering, National Research Council, and Assistant Secretary, National Academy of Engineering

PREFACE

Of increasing concern to many of the nation's leaders in government, business, and organized labor is the gap between the potential technical and entrepreneurial vitality of the U.S. economy and its actual performance. This concern is reflected in related issues, such as the following: Has the rate of technological innovation in the United States been high enough to ensure continued economic growth? Is the so-called "productivity slowdown" of recent years a major problem, and to what extent, if any, is it related to a slowdown in the rate of technological innovation? Have U.S. exports of high-technology goods, considered by many to be indicators of relative U.S. strength in the international economy, been growing at a satisfactory rate? Has the transfer of U.S. technology to foreign nations been detrimental to U.S. national security or to domestic economic performance? The purpose of this report is to explore these issues and to examine their implications for U.S. governmental policy.

   The National Academy of Engineering (NAE) and the National Research Council (NRC) have conducted a number of seminars and studies of the relationships between technology and the economy over the last few years to examine the major issues relating to the complex subject. On October 14 and 15, 1970, the NAE held a symposium on "Technology and International Trade." At its annual meeting on April 24, 1975, the NAE conducted an open seminar on "U.S. Technology and International Trade." Early in 1976 the Department of Commerce and the National Science Foundation requested that the NRC evaluate the subject of "Technology Transfer from Foreign Direct Investment in the United States." This was part of a larger study conducted by the department for the Congress under the Foreign Investment Study Act of 1974 (PL 93-479). The

NRC report dealt with technology transfers actually or potentially arising from foreign investments in the United States in four industries: pharmaceuticals, electronics (including computers and scientific instruments), non-electrical machinery, and petrochemicals and their derivatives. It was published in April 1976 as Volume 9 of the report by the Department of Commerce, *Foreign Direct Investment in the United States*.

In August 1976, the NAE and the Assembly of Engineering of the NRC held a workshop at Woods Hole, Massachusetts, entitled "Identification and Definition of the Technological Factors Contributing to the Nation's Foreign Trade Position and Domestic Socioeconomic Development."

The purpose of this exercise was to explore systematically the important inter-relationships between technological innovation, U.S. export performance, and domestic U.S. economic development with a view to identifying and defining major problem areas. While it was hoped that solutions to at least some of the identifiable problems might be proposed at the workshop, it was recognized that for many problems this would not be easy; for these, further study would be necessary. At the workshop in Woods Hole, consideration was given to U.S. constituencies having a direct interest in these problems and to policy actions pertinent to their interests.

This report is based upon the Woods Hole workshop. It presents those issues on which there was consensus as well as those for which there were conflicting viewpoints. Because various, often strongly opposed constituencies were represented in the workshop, it is hardly surprising that this report, in addition to illuminating the important issues, exposes widely varying points of view toward them. It is clear that the possibility of simultaneously satisfying the often conflicting goals of providing more jobs, improving environmental quality, and supplying better education, health, and welfare for U.S. citizens--while at the same time developing new technologies for conservation of energy and materials--requires a healthy economy and an innovative climate in the United States.

Even during periods when the U.S. economic position is strong, the current and the possible performance of our economy must be examined--in detail and not just in the aggregate--and appropriate actions to improve performance must be taken. Some observers argue that the scientific and engineering capability in the United States is poorly mobilized with respect to economic and social goals and is not as effectively directed toward industrial innovation and productivity as it might be. It is argued that federal investment in research and development is excessively committed to military and space activities to the neglect of civilian technologies or that trained technical people in the United States are not used to best advantage by the private sector. In addition, there are other concerns relating to technology and the health of the economy that bear examination: federal and private support for research and development; the impact of regulatory, antitrust, monetary, and tax policies; industrial structures; and labor "adjustment assistance" programs.

THE WORKSHOP

The purpose of the 1976 workshop was to examine U.S. policy with respect to the technological factors that affect U.S. international trade. The policies examined may have been explicit in their intent to affect trade matters, or promulgated for other reasons but directly influencing international trade.

The workshop was structured to bring the committee together with decision makers from the private sector, including corporate executives, investment bankers, and research managers, as well as officials from a number of federal agencies, professors of economics and law, and labor union leaders. The professional backgrounds of the participants included the sciences, engineering, management, finance, economics, law, and the crafts. Despite this diversity of backgrounds and experience, participants had much in common. Most occupied senior positions of leadership in corporations, universities, and the government. Most had first-hand knowledge of the impacts of technology on U.S. international trade and investment.

In preparation for the workshop, Professors J. Herbert Hollomon and George Heaton, of the Center for Policy Alternatives at the Massachusetts Institute of Technology (MIT), completed *A Factbook Concerning the Relationship Between Technology and Trade*, Volume 1, and *Legal/Institutional Data*, Volume 2. Prior to the workshop, a series of interviews was held in Washington with senior government officials to obtain their views on pressing issues relating to technology and trade. Some of these issues were summed up by Professor E. M. Graham of the MIT Sloan School of Management in a background paper entitled, "Technological Innovation, the 'Technological Gap,' and U.S. Welfare: Some Observations."

These background materials, plus recent, mainly scholarly analyses of the subject, provided the framework for the workshop, which encompassed (a) the factors that influence trade between the United States and foreign countries, (b) how such factors have been changing in past decades, (c) the policies and programs of other countries and the United States relative to the factors and the changes, (d) the important problems and critical issues faced by the United States as a result of the changes, (e) the means of bringing such problems and issues to the attention of decision makers in such a way that they can begin to take actions, and (f) a program of study and analysis that may be useful in initiating relevant actions by the United States. The workshop addressed the following concerns:

- Technology transfer between the United States and other OECD (Western industrialized members of the Organization for Economic Cooperation and Development) countries and the implications of transfer of technology for patterns of U.S. trade with these nations.
- The relationships between technological innovation and U.S. productivity. The effects of exports of technology and capital upon U.S. levels of employment.
- The effects of technology transfer upon the development of the less developed countries and the impact of this transfer upon U.S. trade with these nations.
- Trade and technology exports in relation to national security.

The principal chapters of this report were prepared by Professor E. M. Graham of the MIT Sloan School of Management and were reviewed by the Study Committee and some of the participants who are particularly expert in specific subject areas.

CONTENTS

Abstract of Issues and Recommendations		1
1	Introduction	11
2	A Background Review of the Relationships Between Technological Innovation and the Economy	18
3	Technology Transfer and Trade Between the United States and the Other OECD Nations: Critical Issues	49
4	The International Transfer of Technology, International Trade, and International Investment: The Point of View of U.S. Organized Labor	71
5	U.S. Trade and Technology Transfer to the Soviet Union and the Eastern European Nations	92
6	Technology and Trade Issues Relating to Developing Nations	120
Appendix A: Commentaries on the Change in the Relative Economic Status of the United States With Respect to Other OECD Nations		139
Appendix B: Transfer of Technology to Developing Nations: The Financial Aspects		152
Selected Bibliography		157

ABSTRACT OF ISSUES AND RECOMMENDATIONS

The choice of the designation abstract rather than summary is deliberate. The report itself is in essence a summary--not a study in the usual sense, but an analytical and interpretive summary of 10 days of workshop discussion. The workshop brought together 39 individuals of diverse backgrounds to discuss a broad and complex tangle of issues relating to technology, international trade, and international investment and their implications for the U.S. economy. The discussants came from private industry, organized labor, departments and agencies of the federal government, universities, private research organizations, and, in one instance, an international organization. Their training and experience included the natural sciences, engineering, management, economics, government, law, and the crafts.

The workshop was directed toward possibilities of policy and action, with action defined broadly to include programs of research, and the makers of policy and actors defined broadly to include the U.S. government, private business enterprises, and universities. The discussion was organized in four main sectors reflected in the chapters of the report: chapter 3 deals with issues in technology transfer, trade, and investment among the United States and the other Organization for Economic Cooperation and Development (OECD) nations; chapter 4 concerns the posture of U.S. organized labor in relation to technology transfer and investment abroad by American enterprises; chapter 5 concerns technology transfer, trade, and investment between the United States (and other OECD nations) and the Soviet Union (and other East European nations); and chapter 6 deals with issues of technology transfer, trade, and investment relating to developing nations.

The report, an analytical and interpretive summary,

characterizes the wide-ranging, varied, and intricate substance of the workshop. The substance is amplified with history, data, and theory. This abstract delineates the principal policy issues that emerged, and, with respect to issues on which the workshop participants reached a consensus, it presents the recommendations for policies and actions, with incidental references to the discussion itself.

The issues and recommendations that follow are grouped to correspond to the main sectors of our inquiry. The OECD sector of the discussion exposed not only issues of particular application within the OECD group itself but also issues of general significance and application to U.S. trade with all countries.

ISSUES AND RECOMMENDATIONS OF GENERAL APPLICATION

A tangle of issues relating to technology transfer, technological innovation, and productivity, emerging initially in the discussion of international trade and investment among the OECD countries, engaged the participants throughout the workshop. What have been the trends in technology transfer, technological innovation, and productivity in the United States and abroad? Has the lead of the United States in the development of new commercial technology been diminishing? If it has, does it matter? To the extent that technological innovation may have decreased, is this related to trends in technology transfer? Should the federal government attempt to restrict the transfer of technology abroad by American enterprises, whether to protect the domestic economy or to safeguard the national security? Should the national government seek to provide new incentives for technological innovation, whether through subsidies or by other means? Has the competitive position of the United States in international trade been adversely affected by a lagging rate of improvement in productivity in the United States? Does the current U.S. position in international trade suggest that new measures should be adopted by the federal government, in addition to the policies and programs now in effect, to accelerate the rate of improvement in productivity in the American economy?

Much of the discussion of such questions was based on a tacit postulate that the diffusion of technology abroad by American firms has acquired new significance and created new problems in the past two decades. For whatever reason, technology transfer has become a subject of lively and widespread concern that is shared by the workshop participants. Some members of the group believe that a significant shift has occurred in recent years in the nature of technology disseminated abroad by American firms. In the earlier years, according to these members, American firms typically transferred technologies to other countries in the form of well-established types of products and production techniques that were widely understood. Recently, however, according to this view, American firms have tended to disseminate new and highly sophisticated technology encompassing production systems, process engineering, and management know-how. Even so, most of the participants in the workshop, while recognizing differences in modes of transfer and in types of technology transferred, did not discern any pronounced shift from one dominant mode or type to another from decade to decade.

The discussion covered the possible causes for increases in technology transfer and changes in modes of transfer that may have occurred. The participants pointed to trade barriers imposed by foreign governments and subsidies by some governments to their own enterprises and to U.S. governmental encouragement of technology transfer to East European countries in an effort to reduce their dependence on the Soviet Union. In addition, workshop members discussed the ready availability in some instances of financing from foreign sources (e.g., large Japanese firms) for American enterprises on occasions when the U.S. companies found it hard to come by needed capital within the United States. Attempts by American enterprises to avoid costs arising from regulatory measures or high wages at home; technology transfer into the United States, as well as from it, resulting from direct foreign investment in the United States; and, more generally, "the forces of the market" were also considered.

Of central importance to the workshop were discussions of technological leadership and innovation in the United States and abroad and the implications of technology transfer for employment and the general economic welfare of the United States. Many of the participants believe that the lead of the United States in technological innovation has diminished in many industrial sectors in contrast to some countries in Western Europe and Japan. Some say that the transfer of technology abroad by American enterprises has contributed significantly to eroding American technological leadership. Others attribute the reduced margin of leadership not so much to a decline in the development of U.S. technology as to a surge of technological innovation and productivity in some countries in Europe and Japan. The recent rapid technological advances in those nations were due, in part, to the release of capabilities that had been delayed by World War II and its aftermath and had not come to full fruition until the late 1960's and 1970's. Many participants observe that the often alleged deterioration in the dynamism of the U.S. domestic economy is the primary cause of a reduction in U.S. technological leadership; in their view, the diminished lead in technological innovation is connected with lagging rates of improvement in U.S. productivity. In the discussion, even the concept that it was in the best interests of the United States to maintain its technological lead (in all respects) was brought into question by some participants, who advocated a concentration by the United States in certain high-technology industries.

The search for causes and the attempt to appraise the consequences were related to the primary mission of the workshop: to assess the importance of technological factors on trade and to discover and explore possible remedial policies and actions, if they were called for. While the many differences in approach and analysis among the participants led, inevitably, to differences in proposals for action, the participants did agree on certain suggested measures. The two most important are as follows:

• The U.S. government should not enact legislation to restrain the export of commercial technology by private firms from the United States (apart from possible special measures designed to meet special problems as they arise, for example, in trade with the Soviet Union).

• A general inquiry should be organized into all possible ways and means to foster technological innovation in the United States. This inquiry should range broadly over tax policy and incentives, regulatory policy, antitrust practices, and other federal laws and policies affecting innovation.

Two additional conclusions were specific in nature:

• The U.S. government should adjust its priorities in the allocation of financial support to research and education, by giving enhanced attention to research and education affecting productivity and innovation in commercial technology and by seeking effective ways to provide financial support to research in process engineering and production systems and to imaginative engineering training

related to the processes of production. This adjustment
should and could be made while federal support for basic
research is fully maintained or, indeed, increased.
- An inquiry should be undertaken into the comparative
age and quality of the stocks of capital goods, arranged
by industrial sectors, within the United States and other
countries. Such a study should emphasize comparison of
U.S. industries with those in other countries in the OECD.

ISSUES AND RECOMMENDATIONS CONCERNING TECHNOLOGY TRANSFER,
TRADE, AND INVESTMENT AMONG THE OECD COUNTRIES

The issues and recommendations in the preceding section are
relevant to relations among the United States and the other
OECD nations, as well as to relations of the United States
with countries outside the OECD community. Another set of
issues and recommendations has a special bearing upon technology transfer, trade, and investment within the OECD.
What have been the domestic economic costs and international
competitive implications of certain U.S. regulatory legislation designed to serve such valid and varied purposes as
environmental protection, occupational and personal health
and safety, public participation in decisions of public concern, and protection against undue financial or corporate
concentration? In the enactment of such legislation and
in its administration, have the economic costs and international competitive implications been adequately factored
into the total cost-benefit analysis? The workshop discussants expressed concern over the international competitive consequences of differences in the nature, scope, and
vigor of regulatory policies among the several nations in
the OECD and asked whether the consequences of such policies
give a new thrust for harmonizing economic policies among
the OECD nations. Was it necessary and would it be feasible to consider such a new harmony not only of macroeconomic but also of microeconomic policies?
   The members of the workshop recognize fully the difficulty of devising ways to take hold of these problems or
even defining them with precision. Nevertheless, there is
a general sense among the participants that an effective
way is needed to explore the questions raised. A consensus
emerged that one or more workshops or special study groups
should design programs of research seeking to illuminate
and refine the problems and point to ways to deal with them.

ISSUES AND RECOMMENDATIONS CONCERNING THE EFFECT OF TECHNOLOGY TRANSFER UPON U.S. LABOR

In the view of the workshop participants from organized labor, the transfer of U.S. technology abroad and more especially the transfer of entire plants and production systems has had mounting adverse effects upon employment within the United States. In their view, the provisions for adjustment assistance under existing U.S. trade legislation are woefully inadequate to remedy the unemployment and dislocations that have been attributed to technology transfers to other countries. They raised the issue of direct governmental control over the transfer abroad of U.S. technology by U.S. enterprises. In advocating direct controls, they insist that U.S. labor is not turning protectionist or abandoning its support of legislation to foster international trade. They described several of the possible causes of technology transfer: the barriers to free trade interposed by foreign governments and foreign governmental subsidies to their own enterprises, the flight by some U.S. firms from social and environmental regulatory measures at home to countries where comparable legislation is less or lacking; and the U.S. government's encouragement of the export of technology by American firms for foreign policy reasons wholly unrelated to economic considerations. They also pointed to tax legislation that allegedly facilitates the export of technology. In their view, the "market" has very little to do with the phenomenon. As a consequence, they consider that remedial governmental measures to restrain the export of technology, along with capital and jobs, would be fair, reasonable, and beneficial to the nation.

The members of the workshop generally acknowledged the problem and sympathized with the plight of labor; however, they rejected proposals for additional U.S. governmental restraints upon direct foreign investment and technology transfer by U.S. enterprises. The majority of the participants favored other measures:

- The existing readjustment assistance legislation should be amended to expand the scope of eligibility for assistance by including workers who are laid off because the employer has transferred operations to another country or whose jobs are indirectly jeopardized by imports; to extend the time limits during which benefits would be made available; to simplify procedures; and to expand and strengthen job retraining programs.

- Whenever the federal government may be disposed to sponsor the transfer of advanced U.S. technology for foreign policy reasons (for example, in bilateral agreements with other nations), full account should be taken of domestic economic considerations and consequences, including the possible effects upon domestic employment, before a decision to transfer the technology is reached.
- The United States should adopt and seek to give effect to a principle of "tax neutrality" in regard to direct investment and technology transfer abroad by U.S. firms (i.e., taxation considerations should create neither an incentive nor a disincentive for the U.S. investor to invest abroad). Aware of the complexities and difficulties of carrying out such a policy, the participants believe nevertheless that a sustained effort should be made.

ISSUES AND RECOMMENDATIONS RELATING TO TRADE, INVESTMENT, AND TECHNOLOGY TRANSFER BETWEEN THE UNITED STATES (AND OTHER OECD COUNTRIES) AND THE SOVIET UNION (AND OTHER EAST EUROPEAN COUNTRIES)

Two sets of questions dominated the workshop discussion of technology transfer in trade with the Soviet Union. How can and should U.S. institutions and patterns of trade, investment, and technology transfer be adapted to problems arising from national security considerations? What are the actual and potential costs and benefits for the United States of trade with the Soviet Union?

The discussion ranged through a number of subsidiary questions, with emphasis upon problems relating to governmental controls. What degree of control should be exercised by the federal government on the transfer of technology? What materials, goods, and skills need to be controlled? How appropriate and effective are the mechanisms for such control? What are the likely international implications of attempts to control trade in goods and services available from many nations besides the U.S.?

A consensus was reached by the participants on the importance of coordination among the OECD nations that are major sources of technology for the Soviet Union. Such coordination should cover principles and operational guidelines as well as procedures for assuring compliance with agreed principles and guidelines. There was consensus that there was little point in controlling sales that transferred technology if the same technology is available from non-U.S. sources that place no restrictions on such sales.

The force of this recommendation was reduced somewhat by doubts among the participants on the efficiency of existing and prospective bureaucratic controls and concern for the costs of controls, even if and when they could be made efficient. The same doubts were raised about domestic controls.

There was also agreement that the control of sales of "militarily sensitive technology" should be keyed to "lead time" criteria. There should be a presumption against authorizing sales to the Soviet Union of militarily sensitive technologies in which the United States holds a significant lead time advantage. Conversely, there would be little point in restricting sales to the Soviet Union of technologies in which the United States has only a negligible lead time advantage.

The participants stress the need for more insight into problems of accommodation between the respective institutions of the United States and the Soviet Union that determine patterns for trade, investment, and technology transfer. In the members' view, there has been a tendency among OECD governments, business enterprises, and scholars to pass too lightly over the institutional differences and their possible consequences. As trade increases between the United States (and OECD nations generally) and the Soviet Union, the United States (and other OECD countries) might be confronted with a choice between serious impairment of their own comparatively open trading and payment systems and a blunting and reversal of the expansion of trade with the Soviet Union. These participants had no solutions to offer to the problem they defined, but they urge that more attention be given to the problem.

ISSUES AND RECOMMENDATIONS CONCERNING TECHNOLOGY IN DEVELOPING COUNTRIES

The discussion of technology transfer to developing countries centered upon the demands of the Group of 77* for modes of technology transfer appropriate to the New International Economic Order. In a sense, this added a new

---

*Historically, the Group of 77 is that informal association of developing nations which came into existence at the 1964 United Nations Conference on Trade and Development (UNCTAD) and pressed the industrialized nations for a new international economic order. The Group of 77 now comprises 111 member countries of the United Nations.

thrust to the workshop inquiry, as the scope of the workshop was enlarged from an assessment of harms and benefits to the United States to an assessment of benefits and harms to the developing countries as perceived and articulated by their spokesmen, notably the Group of 77. In another sense, however, the workshop maintained its original direction and emphasis but interpreted "effects upon the United States" in terms of the nation's foreign policy as well as its foreign trade position and domestic economy.

Examination of this issue began with a review of arguments voiced by the Group of 77 against present technology transfers--that the cost of technology transferred to the developing countries through direct foreign investment by multinational corporations is unwarrantedly high; that the technology transferred is not suited to the special capacities and needs of the recipients; and that multinational corporations engage in unfair practices that discourage the rise and development of local enterprises. Many of the workshop participants believe the assertions to be unfounded or exaggerated and significant primarily as reflections of the Group of 77's aspirations and frustrations. The discussion turned to possible ways and means by which the United States might accommodate its policies in some measure toward relief of the frustrations and support for the aspirations.

In the end, the workshop recommended that

- A program be established by the National Science Foundation or other appropriate government agencies to carry out a continuing analysis of ways and means by which the United States could respond to the desire of the developing countries for technology appropriate to their needs and capacities.
- The United States consider plans to assist financially the establishment within developing countries of regional institutions of applied research and development (R&D), as well as the creation of international institutions designed to contribute to understanding of the industrialization and economic development process in developing countries. Several participants warned that the usefulness of regional institutions would depend on how effectively the work could be assimilated into the outlook and practice of local enterprises that were expected to use the R&D results in production.
- The United States explore possible new programs (or seek to improve existing programs) that would (1) assemble and maintain inventories of technologies that the federal

government owns or has legal capacity to transfer and (2) facilitate the selective transfer of such technologies to developing countries.

- The United States cooperate with other OECD nations in coordinated programs for the selective transfer to developing countries of technologies that OECD governments own or have legal capacity to transfer.

1 INTRODUCTION

The Woods Hole Workshop consisted of 10 days of discussion among 39 individuals whose professional training and occupational endeavors made them uniquely qualified to present points of view representative of major constituencies in the U.S. economy.

Among the policy issues considered by the group were the following: Should the U.S. government attempt to restrict the transfer of technology abroad, either to protect the domestic economy or for national security reasons? Should the Congress enact laws to provide new incentives or subsidies for technological innovation? Would measures by the federal government to increase the productivity of the domestic economy (or otherwise to provide for more efficient use of existing domestic resources) improve the competitive position of the United States in international trade? Should the United States support or initiate increased efforts to transfer technology to the developing nations?

This report provides the essence of the discussion of such questions.

SOME KEY CONCEPTS

It is essential to any rational discourse that the discussants have a common understanding of what is being discussed. Cognizant of this, the workshop participants endeavored to define precisely the topics that they sought to cover.

The key theme of the workshop was American technology and its role in determining the position of the United States in the international economic system. However, to define technology is not unequivocably easy. Technology is a type of knowledge, the "know-how" necessary for the creation of goods and services demanded by economic society. However, there is no agreement on exactly what types of knowledge are to be included under the rubric of technology. To some, all economic knowledge is technology, including the know-how associated with marketing, finance, and other managerial functions, as well as purely technical know-how. Others consider technology to be limited to technical know-how, i.e., knowledge pertaining specifically to the design, manufacture, and use of manufactured or processed goods. While the workshop participants, for the most part, accepted the second, more restrictive definition, the boundary between what types of knowledge are or are not technology was not strictly defined.

Several important points were made with respect to what technology is *not*. First, technology is knowledge, not goods and services. Goods and services may embody technology, but they are not technology in and of themselves. (For example, an advanced computer system and its software are not in themselves technology, although they may embody a great deal of technology.) Second, technology is not synonymous with scientific knowledge. Scientific knowledge is generally recognized to be civilization's cumulative understanding of the functioning of the universe. Although technology is based upon an application of this understanding, the application is often predicated upon the accumulation and use of additional knowledge quite distinctly different from the basic scientific knowledge that underlies the application. This additional knowledge may frequently be relatively mundane, determined perhaps more by trial and error than by deductive reasoning, but it is as essential a component of technology as is the underlying scientific knowledge.

Technological innovation is a term that has come to be distinguished from the invention of a new technology. Technological innovation is defined as the total process by which new technology[1] is generated and carried through development to the point where commercially usable products (or services) are introduced to the marketplace. Technological innovation is a complex process, one that was discussed at some length by the workshop. The complexity of the innovative process reflects the economic difference between the creation of scientific knowledge and the application of this knowledge to the advancement of economic welfare. It was noted by the workshop that when discovery of new scientific knowledge leads to technological innovation, the intervening processes are often long, complex, and costly. Much of the cost and time are associated with the stages beyond the generation of the basic technology itself, specifically, with the production and marketing of new products made possible by new technology. An example given was that of the innovation of the modern drug penicillin. The scientific discovery that led to the development of penicillin, as most school children learn, was Alexander Fleming's (semifortuitous) discovery in 1928 that the mold *Penicillium notatum* emits a substance that is strongly antibacterial but relatively nontoxic to the mold itself. From this, school children are taught to infer (incorrectly) that Fleming "discovered" penicillin in 1928. The fact is that the substance we now call penicillin was not even isolated until about 1938, or about a decade after Fleming's discovery. Several dozen persons (including Fleming) devoted many dozens of man-years to the task of accomplishing the isolation. Even after the substance was first isolated, tens of millions of dollars and hundreds of man-years additionally had to be invested before the development of a clinically useful drug was achieved, and, even under accelerated development schedules occasioned by World War II, large-scale introduction of the drug was not accomplished until 1944. Vast improvements both in the efficacy of the drug and the efficiency of its manufacture took place over the next two and a half decades.

In the case of penicillin, the most costly and time-consuming element in the development of the drug was not the basic scientific research but rather the painstaking development of very specific, often mundane, procedures and methods for producing a standardized drug in large batches. The innovation of penicillin illustrates a characteristic of most technological innovation: the relative cost of the development phase of R&D is greater than the cost of the research phase.

If the case of penicillin were unique, there would be little or no point to the preceding narrative. The innovation of penicillin is not, however, at all unique. Technological innovation is in almost all cases an evolutionary and costly process. Historians generally date the invention of the internal combustion engine, for example, to somewhere around 1885, but the innovative process associated with the development of this engine has never really ceased--in fact, under the impetus to make the operation of the engine more efficient and clean, one might hope that much important innovation is yet to come.

Technology is often disaggregated into "product" technology and "process" technology, the latter usually being defined as the knowledge embodied in the means by which a product is made or manufactured, while the former is defined as the knowledge embodied in the design of the product itself. This disaggregation is a bit artificial because what is at one level a product might be at some other level a process. A machine tool, for example, is very definitely a product, but it is a product that is destined to be used as a part of a manufacturing process for some other firm. This same product/process duality applies to many products and processes--even household goods such as home appliances are part of a process in that they contribute to the production of goods and services in the home. Accepting that product technology and process technology are not strictly separable concepts, the workshop nonetheless found it useful in many cases to think of them as being distinct from one another at a given level of production. Thus, for example, in the chemical industry, one might talk about product technology as the knowledge associated with the chemicals themselves and process technology as the knowledge associated with the plant and equipment used to make the chemicals.

A major concern of the workshop was whether or not the price at which U.S. technology is transferred abroad is reflective of the true economic value (to the United States) of the technology. While this issue is discussed at some length in later chapters, a number of preliminary observations are offered here.

The question of the valuation of technology (or any other form of knowledge) is a complex one. Once technology is created, it can be utilized by all individuals within an economic society at a zero social cost (other than costs associated with learning). This zero social cost of utilization of existing technology has, in fact, led neoclassical economists to reason that maximum benefit from existing technology would be achieved if it were made available to all potential users free of charge.[2]

The problem, of course, is that the creation of technology requires an investment of real resources. In a private market economy, this investment is borne by the creator of the technology. Unless adequately compensated for this investment, a private firm has no incentive to generate new technology. This problem is handled by allowing the creator of the technology to hold a temporary monopoly over its utilization, the monopoly being legally embodied in a patent. The patent allows a firm which has created a new technology to control its commercial exploitation, either by using the technology solely within the firm itself or by allowing other potential users to utilize the technology for a fee. Because this control results in restriction on the exploitation of knowledge, short-run economic efficiency may not be maximized, but this possible inefficiency is accepted as the price society must pay for the existence of a continuing incentive among private firms to create and publish new technology. The patent serves to disseminate knowledge more broadly, through its publication, and thereby may stimulate further invention elsewhere. Thus its net effect may be to increase economic efficiency in the long run.

It has been argued that the patent system does not allow the creator of new technology to capture the full potential economic value of the technology. This comes about for at least two reasons. First, any patent is temporary, and thus the time over which a firm can exploit the patent is limited. Second, it is possible for a firm's competitors partially to circumvent a patent by developing and utilizing technologies that are not covered by the patent but that compete directly with the patented technology; the primary advantage given by the patent is often the lead time it furnishes. Also, the full social value of the technology would be realized only if its use were complete. To capture any private return to itself, the holder of the patent must restrict its use by competitors either by charging a fee for its use or withholding it from competitors. This does not necessarily result in a reduction of the social value achieved, of course, as the ultimate users can still have full access to the new technology and its benefits. Some economists have been led to suggest that under a patent system there is likely to be an underinvestment in the creation of new technology, an underinvestment that can be compensated for only by providing some additional public incentive for such creation by private enterprise beyond that afforded by the patent (this point is further discussed in Chapter 3 of this report). This conclusion, however, is not universally accepted by all economists.* Industry is quite generally in favor of a strong patent system as an incentive to investment in new technology.

The above analysis leads to a major dilemma when one attempts to assess what is the "value" of technology transferred abroad. The nature of the dilemma is illustrated

---

*See conflicting views presented by several economists in *Relationships Between R&D and Economic Growth/Productivity*, Preliminary Report (Washington, D.C.: National Science Foundation, 1977).

by means of considering the cost and value of the transfer of technology from the point of view of both the receiving and the donating nations.[3] To the receiving nation, the value of the transfer of technology is equal to the value of the social benefits that result from its usage. The total cost it would be willing to pay to have the technology transferred is anything up to this amount. To the donating nation, the cost of the transfer of technology is a function of the extent to which this transfer affects the rate of investment in the creation of new technology. If international transfer of technology has no effect upon the rate of investment in domestic technological innovation, the cost of transfer is zero; after all, the transfer of technology abroad does not reduce the availability of its use at home. In principle, in this case, the donating nation should be willing to transfer its technology free of charge!

The fact is, however, that international transfer of technology does affect the rate of domestic investment in technology in a private market economy. If a technology created by a domestic firm is transferred, say, to a foreign competitor of that firm without compensation, the portion of the social value of the technology that can be captured by the firm is reduced. Accordingly, the incentive to create new technology is also reduced. Thus, assuming that the net social value of new technology is positive, there is a social opportunity cost associated with such a technology transfer.

This opportunity cost can be compensated for in a number of ways. If the technology is transferred by a multinational corporation from its home nation to a subsidiary within a foreign nation, the firm will be able to earn some private return from utilizing the technology in the foreign market. This private return might then be reinvested in the development of new technology that could be utilized in both the home nation and the foreign nation market. This is a point that advocates of the multinational corporation rightly stress--that the multinational corporation is an effective institutional mechanism for spreading both the benefits and the costs of technological innovation. Compensation of social opportunity costs can be accomplished by other means, however. The two principal such means would include licensing, which returns to the licensor a fee for utilization of technology, and a lump-sum payment for technology transfer.

If the social opportunity costs of technology transfer for the donator nation are less than the social value of the transfer to the receiving nation, it presumably would be in the interest of both nations that the transfer take place. The price of the transaction presumably would fall between the social cost to the donor nation and the social value to the receiving nation. Exactly where it would fall, however, is not analytically determinable and probably depends to some extent upon the relative bargaining capabilities of the two parties.

NOTES

1. Or a previously unused "new" technology.
2. For an elaboration of this discussion, see H. G. Johnson, "The Efficiency and Welfare Implications of the International Corporation," in C. P. Kindleberger, editor, *The International Corporation* (Cambridge, Mass.: MIT Press, 1970).
3. In the following discussion, the benefits accruing to both nations from increased specialization of production resulting from technology transfer are deliberately ignored.

2 A BACKGROUND REVIEW OF THE RELATIONSHIPS BETWEEN TECHNOLOGICAL INNOVATION AND THE ECONOMY

This second chapter is a brief introduction to some of the major thought that has been advanced relating technology to the economy. This material is presented as background for the remaining chapters of the report, which deal specifically with the workshop discussions. The material reviewed here was selected on the basis of what is most relevant to these discussions. By and large, the review emphasizes empirical studies over more theoretical approaches to the study of the economic role of technological change. The reader who is familiar with the literature may wish to skip this chapter and proceed directly to Chapter 3.

1. EFFECTS OF ADVANCES OF TECHNOLOGY UPON THE ECONOMY: SOME PRELIMINARY OBSERVATIONS

This section briefly explores how technological change affects the economy, at least as it is understood within a neoclassical economic framework. One word of caution is in order. Most economists would certainly agree that technology is a major factor in the economic process and that technological innovation is an important source of economic growth. Exactly how technological innovation leads to growth is not fully understood, however, and what understanding does exist is difficult to capture in theoretical models of the economy.

Under neoclassical economic thinking, the major economic effect of technological innovation is savings in factor usage. Simply stated, the effect is to enable the production of goods to be accomplished with less total input of resources, that is, to enable the goods to be produced more efficiently. More efficient production of goods, in a competitive economy, results in a lower price of these goods. The lower price, in turn, leads to expanded demand and expanded output of the good and thus to economic growth. This growth is made possible because resources that previously were expended to produce one unit of the good are now released for use elsewhere in the economy.

Technological innovation that results in increased efficiency is classified by economists as being "factor neutral" or "factor biased."[1] An innovation is factor neutral if, at constant (preinnovation) relative prices of inputs and constant levels of physical output of the good, adoption of the innovation results in no change in the relative quantities of inputs consumed per unit of output. The innovation is said to be "factor biased" if, at constant relative prices of inputs and constant levels of physical output, there is some change in the relative quantities of inputs consumed per unit of output. For example, suppose that a technological innovation in some industry creates a new method to produce the product of that industry more cheaply. There is no change in either industry wage levels or producers' cost of capital. Producers who adopt the innovation do not increase (or decrease) output, but they replace workers with machinery. The innovation in this case would then be classified as "capital using" or "factor biased in favor of capital." Later in this chapter it will be noted that certain hypotheses regarding the role of technology in U.S. foreign trade implicitly assume that technological innovation in the United States is largely capital using.[2]

It is evident that if technological innovation is capital using, the substitution of machinery for labor will cause workers to become unemployed, in the short run at least. However, in the longer run, two effects will correct for this short-run trend. First, the supply of labor will increase in relation to that of capital, causing the relative price of labor to capital to fall.[3] Consequently, labor will be substituted for capital, and full levels of employment will be maintained in the long run. Second, economic growth resulting from the technological change will provide new, hitherto nonexistent employment opportunities for labor displaced by the change. Typically, as the economy grows, the new jobs created will be more productive than the old jobs eliminated, and thus the general levels of wages will rise. On the whole, then, in the long run, technological change benefits both the economy and the worker. The economy and the labor force, however, might experience certain problems adjusting to a changing technology. In particular, it has been argued that if an economy produces a constant stream of (capital-using) technological innovation, there will be a persistent lag between adoption of the new technology and adjustment.[4] One consequence of this lag might be some sort of persistent unemployment.

It has been argued that capital-using technological innovation might have the effect of changing the qualitative component of demand for labor. Capital-intensive production processes might require, for example, more skilled labor than would labor-intensive processes. Thus, the long-run effect of capital-using innovation might be to increase the demand for skilled labor but to reduce the demand for unskilled labor.[5]

Most formal analysis of the effects of technological change upon the economy centers upon increases in efficiency as the major such effect. Certainly, an increase in efficiency is a major consequence of innovation in new process technology. More difficult to gauge than innovation in process technology is the economic effect of innovation in product technology. Such innovation often results in the creation of products that before did not exist. A successful introduction of a new product has at least three possible effects on the economy, and these effects are complex and interrelated: (1) Consumers' tastes may change, creating a demand for the new product that did not exist before the innovation. The new product might, additionally, displace older, imperfectly substitutable products. The net result is a change in the product mix of the economy. (2) There is created a new demand for inputs to produce the new product, resulting in new investment and new employment opportunities. However, it must also be considered that there may result reduced demand for production inputs for the displaced products, so that the net change in demand can be either positive or negative. (3) The new product might itself be used as part of the production process in some other industry, and thus contribute to increases in efficiency in that industry.

The full economic consequences of these three potential effects of new product innovation are not fully understood. In passing, it might be noted that because the only effect of technological change that can be readily embodied in economic models is that of increased efficiency, most models of the economy that attempt to capture the technological factor limit themselves to increases in welfare resulting from increases in efficiency. This would probably not be a bad approximation if the majority of technological innovation were of the "process" variety. Empirical studies, however, have indicated that over 40 percent of expenditure by U.S. industry on research and development is devoted to the creation of new products and almost 45 percent is devoted to product improvement, leaving only about 15 percent for direct expenditure on process improvement.[6] It is probably safe to say that the effect of technological innovation upon the U.S. economy is more profound than simply increases in efficiency. In particular, qualitative changes in the product mix through the introduction into the economy of better consumer products doubtlessly raises the level of economic welfare, but the magnitude of the increase is not readily measurable.[7]

2. THE ECONOMIC BASIS FOR TECHNOLOGICAL INNOVATION

The previous section briefly discusses some of the principal effects of changes in technology upon the economy. This section takes up the topic of what induces technological change to occur within an economy.

Perhaps historically no one author has had quite as much impact upon thinking about this topic as Joseph Schumpeter.[8] Schumpeter was primarily concerned with the effect of the introduction of new technology upon the business cycle, but to explain the business cycle, he devised a theory of why technological change occurs in an economy. His theory sharply distinguished among technological invention, innovation, and imitation. Technological invention, as seen by Schumpeter, was a more or less continuous process that occurred outside of the mainstream of economic activity but that created a pool of new technology (i.e., knowledge embodied in new products or processes) that could at any time be tapped by the economy. Schumpeter never really explained (or was concerned with) the process by which this invention occurred.[9]

Technological innovation, according to Schumpeter, occurred when the entrepreneur singled out certain inventions and commercialized them. Specifically, innovation was considered to be a change in the production function brought on by any of five cases: (1) an introduction of a new product, (2) an introduction of a new process, (3) the opening of a new market, (4) the discovery of a new source of raw materials, or (5) the reshaping of the market structure of an industry. (Today, we would probably encompass in the definition of technological innovation only the first two of these.) The essence of Schumpeter's theory was that it was only at irregular intervals that innovation took place, so that innovations, over time, occurred in clusters. The reasons for this clustering had to do with both the resistance of society to change and with the visionary, leadership role of the entrepreneur. Once the entrepreneur had exerted the leadership required to innovate and thus had shown the way, it would be relatively easy for lesser souls to imitate the innovation or to make similar or additional innovations of an incremental nature. Thus, a major innovation followed by a flurry of imitation actually accounted for the clustering.

The clustering of innovations was to Schumpeter at the heart of the business cycle. The appearance of a cluster of innovations would give impetus to the economy to grow. However, because the innovations were not continuously forthcoming, the impetus would be short-lived. Without the impetus, the growth rate of the economy would eventually stagnate or go into decline, until a new round of innovative activity would come around to generate a new cycle of growth. Schumpeter's concepts, it might be noted, stand in contrast to the Keynesian explanation of business cycles.

Criticism of Schumpeter has been focused upon the very sharp distinction he placed between invention, innovation, and imitation.[10] It is argued that these activities may not be nearly as sharply differentiated as Schumpeter thought, and that technological innovation might in fact be much more an incremental process than Schumpeter described it. Invention, it is argued, does not necessarily occur in an economic vacuum; rather, economic forces have significant impact upon the types and magnitudes of inventive activity that take place.[11] Likewise, the process of imitation might not be as secondary to the innovation process as Schumpeter thought. If, in fact, the innovative process is an incremental one, the ultimate form that the innovation takes might be highly dependent upon the manner in which the end product of the innovation diffuses into economic society. Imitation, in this context, is more than riding the coattails of the innovator. The success of the imitator might be highly dependent upon his ability to effect improvements in the product or to embody in the product those changes as demanded by the marketplace.[12]

Schumpeter's empirical observation of clustering of innovation is challenged largely on the grounds that he considered only major, quasi-revolutionary technological innovation and in doing so ignored many smaller contributions to technological change. The irony of this exclusion is that in many cases the small, evolutionary advances in technology were prerequisite for the major innovation to occur. As an example, the diesel engine, conceptually "invented" in the late nineteenth century, was not really commercially "innovated" until after World War II. A practical, economically feasible realization of Rudolf Diesel's ideas had to await advances in metallurgy and petroleum refining before, in a Schumpeterian sense, an invention could be transformed into an innovation.

The lag time between invention and innovation is a subject of a classic study by John Enos, in which it is reported that for a sample of 46 innovations there was an average interval of 11 years between invention and innovation for the petroleum industry (Enos' study focused on the petroleum industry) and about 14 years for other industries.[13] The variance in these figures is very large, ranging from a maximum of 79 years (the fluorescent lamp) to a minimum of just 1 year for freon refrigerant. The specific findings of Enos are criticized by Nathan Rosenberg, who suggests that Enos' approach to the identification of the date of invention of products might be spurious.[14] Enos defines the date of invention to be "the earliest conception of the product in substantially its commercial form" and the date of innovation to be that of "the first commercial application or sale." Rosenberg suggests that "to date these inventions from an initial basic conceptualization...is to repeat the unwarranted practice...of downgrading engineering and technological forms of knowledge."

Citing numerous studies, Rosenberg argues heavily that invention and innovation are incremental processes essentially indistinguishable from one another. According to Rosenberg, the lag between the demonstration of the technical feasibility of a product or process and its commercialization is accounted for by technical factors and economic factors working in tandem. Examples cited by Rosenberg are the high-pressure steam engine, polyethylene, and synthetic rubber. In the case of the steam engine, early, working, high-pressure steam engines were simply too heavy per unit of power generated to be useful as a ship's power plant. Commercial application of the invention had to await advances in metallurgy and fabrication techniques that allowed the weight to be reduced. Commercial introduction of polyethylene was delayed by problems associated with scaling up the production process. Polyethylene was available in test tube lot sizes years before it was technically feasible to produce it in large batches, but as long as it could be produced only in small batches, its high cost prohibited a successful commercial introduction.

The steam engine and polyethylene illustrate examples of technical problems of an incremental nature standing in the way of economic application of inventions. Before the inventions could be economically utilized, numerous relatively small but collectively critical technical problems had to be overcome.

The example of synthetic rubber illustrates a case in which economic forces forestalled the technical development of a product. Synthetic rubber was known to be technically feasible prior to World War I, but the availability of natural rubber at low cost precluded the possibility of commercialization of synthetic rubber. Only when Japanese occupation of the large rubber plantations of Southeast Asia caused a drastic reduction in the supply of natural rubber did work progress to produce the synthetic variety on a large scale.

The work of Rosenberg reinforces a critical point: the creation of technology (and especially that component of technology that is labeled as know-how) is a time-consuming and costly process. Rosenberg maintains that the neoclassical economist's description of the supply side of the economy, which employs a homogenous production function, in which the factors of production are smoothly substitutable for one another, is an abstraction that deviates from reality because of the implicit assumption that the know-how required to make the substitution is either known or readily available. In most industries in actual fact, Rosenberg argues, the know-how required to achieve factor substitution is simply not available, and hence factor substitution cannot be achieved without investment in research and development.

An important study of the relationship between inventive activity and economic activity was published by Jacob Schmookler in 1966.[15] The main question addressed by Schmookler was as follows: Are inventions mainly knowledge induced (as implicitly assumed by the Schumpeterian framework) or are they mainly demand induced? Schmookler recognized that knowledge is an important determinant of invention, limiting what could be invented and (as new knowledge became available) making possible new invention: "Every invention is (a) a new combination of (b) pre-existing knowledge which (c) satisfies some want."[16] However, after examining a great deal of data on patented inventions and attempting to determine what variables could best explain time series patent data, Schmookler concluded that the amount of invention in capital goods in 20 major industries was largely a function of market variables in those industries. Alternatively, Schmookler concluded that the market determined the rate of invention rather than that the rate of invention determined the market. This conclusion was based on the observation that trends in inventive activity tended to lag trends in investment over time in the industries studied, and that major changes in the trends occurred in the investment time series before these changes occurred in the rate of inventive activity. Furthermore, cross-sectional data indicated that inter-industry differences in the number of capital goods inventions "tend to be proportional to corresponding differences in the number of capital goods sales in the immediately preceding period."

Schmookler's empirical findings were limited largely to inventions in capital goods. Furthermore, his data were largely for industries that had long investment histories, and thus the study did not include modern "high-technology" industries (with the exception of the petrochemical industry, included as part of the petroleum industry). Whether or not similar results hold for consumer goods or for capital goods for very new industries Schmookler was unprepared to state, although he suspected that the answer would be in the affirmative.

It is significant that what Schmookler demonstrated was that the demand for new inventions is a prime determinant of the rate at which these inventions will be forthcoming and thus that an economic theory of technological innovation cannot simply assume an exogenous supply of invention. It would be reading too much into Schmookler's work to say that he saw demand as the only determinant of technological change. He recognized firmly that existing technology, and beyond that, existing science, are determinants of what can be invented at any particular point in time. Existing technology and market demand, to use his own analogy, are rather like the blades of a scissors: both must operate if there is to be invention. Invention (which, to Schmookler, was tantamount to innovation) involved a coupling of market demand to the existing level of knowledge.

3. MEASURING THE IMPACT OF TECHNOLOGICAL CHANGE UPON ECONOMIC GROWTH

To Schmookler, the accumulation of intellectual capital (which, according to him, "of course is but another term for technological capacity"), rather than additions to the tangible factors of production (i.e., land, labor, and capital), is the prime determinant of long-term economic growth. Historically, this has also been the belief of a number of economists, including Karl Marx, Thorstein Veblen, John Stuart Mill, Joseph Schumpeter, and Simon Kuznets. The mainstream of economic thought, in the tradition of Adam Smith and David Ricardo, has historically ignored the role of intellectual capital, at least until quite recently. Beginning in the 1950's or so, however, the role of technology in long-term economic growth became a much discussed issue. The exact contribution of each of the factors of production (including technology) to economic growth has been the focus of numerous empirical studies of recent vintage.[17]

Efforts to measure the contribution of progress in technology to the economic growth rate inevitably involve the measurement of some residual component of growth, that is, the component of growth that cannot statistically be accounted for by measurable increases in tangible factor inputs (land, labor, capital, etc.).[18] This "residual" is attributed to be the increase in output per unit of input, an increase presumably brought about, at least in part, by technological change (but also by increasing returns to scale or more efficient allocation of resources).[19] The list of inputs, if complete, would presumably result in a zero residual, all effects of technological change being subsumed in the inputs. Thus, for example, E. F. Denison, in attempting to measure the differential effects of various inputs upon the growth rates of the United States and Western Europe, separates the contribution of growth of the labor factor into four components: employment (total), hours of work per employee, age-sex composition, and education.[20] These components are not strictly independent of one another, and it is clear that some effects of technological progress could be subsumed in both the age-sex composition component and the education component.

In attempting to isolate the determinants of growth in the United States and Western Europe from 1950 to 1962, Denison considered a total of nine categories of factors and seven categories of output advance per unit of input (including a residual). He concluded that "advances in knowledge" (the residual) accounted for a compound annual growth of 0.76 percent in the United States economy (out of a total annual compound growth of 3.32 percent). In Northwestern Europe, the residual was 1.32 percent out of a total annual compound growth rate of 4.78 percent. Denison assumed that "advances in knowledge" accounted for about the same rate of growth in the United States and Western Europe but that the difference in the residuals was a result of "changes in the lag in the application of knowledge, general efficiency, and errors and omissions."

The methodology of Denison and others attempting to measure the effect of advances in technology upon economic growth is not immune to criticism. In particular, the issue has been raised as to whether such measurements might understate or overstate the contribution of technological change to economic growth. Basing their objections on standard techniques for measuring capital stock, Dale Jorgenson and Zvi Griliches argue that most published measurements tend to overestimate this contribution.[21] Examining the issues in a somewhat different light, however, F. M. Scherer argues that Denison understates the contribution of technological change to economic welfare on the grounds that qualitative improvements in new consumer goods are ignored.[22] The effect of technological change upon economic growth would most certainly be underestimated if technological change were to be factor biased in favor of capital. If so, technological change would induce both economic growth and an increase in the stock of capital. Although the growth would be caused by the technological change (and not by the associated increase in the capital stock), the estimation would (incorrectly) interpret at least some of the growth as resulting from the increase in the capital stock.

An approach to the measurement of the effect of technological innovation upon the economy quite different from that of Denison has been presented by Edwin Mansfield *et al*.[23] The approach is to attempt to measure directly private and social rates of return accruing to successful technological innovation. The private rate of return is calculated on the basis of the net profit of the innovator minus the foregone profits of products displaced by the innovation minus total known research and development costs expended on attempting to develop innovations similar to the final, successful innovation. The social rate of return is calculated on the basis of the above data plus cost savings from use of the innovation accruing to users plus additional demand generated as a result of either a reduction in price of the innovation or reductions in price of end products using the innovation as an intermediate good.

Of the 17 innovations studied by Mansfield, 1 yielded a negative rate of return on both a social and private basis, but the remaining 16 yielded private rates of return (before taxes) ranging from 4 percent to 214 percent and social rates of return from 13 percent to 307 percent. The average private rate of return was about 25 percent before taxes, which the authors felt "does not seem unusually high when one considers the riskiness of this type of investment." (The authors do not, however, present evidence to support their contention that investment in new technology is unusually risky.)

Mansfield *et al*. reported that for about 30 percent of the cases studied the private rate of return was too low for the firm, had it *a priori* known this rate, to have invested in the innovation but the social rate of return was sufficiently high to justify the investment from society's point of view. This conclusion, however, apparently reflected the judgment of the researchers rather than any established normative criteria for "cutoff" or "required" rates of return on investment in innovative undertakings.

In the same paper, the results of a study of the long-term rate of return on research and development activities of one major, unnamed industrial firm were reported. The average rate of return to this firm on research and development for the years 1960-1972 was 19 percent, and it was estimated by the researchers that the social rate of return was much higher. The yearly variance in the rate of return to the firm was high, but no long-term trend in the rate of return could be discerned.

4. EFFECTS OF INDUSTRY STRUCTURE, FIRM SIZE, AND GOVERNMENT FUNDING OF R&D UPON TECHNOLOGICAL INNOVATION

The preceding discussion has ignored the effects of industry structure and firm size upon the rate of innovation. Furthermore, it has not considered that innovative effort might be funded or subsidized by governmental agencies or other exogenous sources. This section briefly touches upon these considerations.

In a 1965 article, F. M. Scherer presented empirical evidence that large firms, as might be expected, contribute a larger portion of technological innovation to the U.S. economy in the aggregate than do small firms but that the proportion of large firms' contributions relative to their total size was less than that of smaller firms.[24] Thus, according to Scherer, smaller firms contribute more innovations per dollar of sales than do large firms. In a later work, Scherer argued that large firms have advantages over small firms in terms of capabilities to innovate, the advantages being greater total resources, economies of scale in R&D, a wider range of corporate activities (one manifestation of which is greater product diversification), and, in principle at least, less risk averseness.[25] Offsetting this somewhat, large firms, according to Scherer, often operate in highly oligopolistic industries in which there exist established patterns of firm conduct that might lead to lower propensity to innovate.

A number of other studies have addressed the issue of whether large firms or small firms are the most innovative. A 1969 study by Jewkes *et al.* suggested that in a number of industries--instrumentation, electronics, and sophisticated machinery--there is a much higher propensity for smaller firms to innovate than for larger ones.[26] The study was based largely upon the innovative efforts of such firms as those along Route 128 near Boston and the area surrounding Stanford University. However, it is not clear that the sample of small firms studied by Jewkes *et al.* are representative of all small firms. Many of the small firms examined in this study carried out innovative activities that were directly or indirectly governmentally funded. Furthermore, most of these firms operate in what might be termed high technology industries, industries in which technology is the main product. It would be doubtful that small firms in, say, the textile or metal-fabricating industries have the same propensity to innovate as those operating in the advanced instrumentation or electronics industries. Many of these small firms were, in fact, founded by entrepreneurs who moved out of larger firms, bringing technology advances from the larger firms with them. Case studies have shown that in many situations, technological innovation will be started by a large firm but for various reasons may not be carried through to completion. An employee of the larger firm can often found his own firm, complete the innovation, and do quite well for himself.[27]

It is in fact very difficult to generalize on whether large or small firms are the most innovative. Without question, certain large firms in the fields of computers, telecommunications, electronics, and pharmaceuticals have been highly innovative, while others in textiles, steel, and shoes have been remarkably noninnovative. Among small firms, while it is doubtlessly true that the Route 128 and Santa Clara County firms have been highly innovative, it is doubtful that the same could be said for all small firms. Two studies of the radio industry, for example, independently reached the conclusion that small firms did little to advance technology during the 1920's and 1930's but that significant advances were made by the largest firms in the industry.[28] On the other hand, studies of other industries have shown the large firms to be the laggards in innovative activity, the brunt of innovation having been carried out by smaller firms.[29]

A distinction too must be drawn between firms that are inventive and those that innovate. In some cases, for example, the process of successful innovation begins with a small company's invention but is carried out to completion by a larger firm. A not infrequent occurrence is for a smaller firm to invest in a product but to run into difficulties attempting to develop it to the point at which it can successfully be marketed. One outcome is for the small company to be bought out by a larger one, which has the resources to complete the process of innovation. On the other hand, it has been already noted that the reverse process also occurs: innovation begins with invention in a large company but is carried out to completion by a smaller firm, often one founded by an exemployee of the larger company.

In some industries, a minimum firm size may be prerequisite for innovation to occur. This is probably true in many sectors of the chemical industry, the pharmaceutical industry, and the heavy machinery industry. In industries such as these, it is argued that economies of scale in R&D are very important and thus that the small corporation has little chance to carry out successful innovation. In such industries, however, a major factor in the introduction of innovation into the industry can be the new entrance of a major firm with a new idea.[30]

A point that has been stressed in many studies of the relationship between industry characteristics and propensity for firms within the industry to innovate has been the role of the new entrant. In oligopolistic industries, wherein patterns of firm conduct are well established, the propensity is often not to innovate until an "outsider" firm seeking new entry into the industry threatens to disrupt the stability of existing intraindustry relationships. The threat posed by the outsider can take many forms: it might reside in some new technology possessed by the outsider but not by established firms, or it might be the willingness of the outsider to cut price in order to build market share, for example. If the perceived threat is in the form of a new technology, the existence of the threat might induce firms established in the industry to create their own new technology, and thus a wave of innovation might occur. In the extreme case, patterns of conduct that formerly discouraged innovation might be abandoned, and new, rivalistic patterns might emerge. The net result might be that the shock of the new entrant induces a flow of innovation.[31]

Consideration of the effects of government funding of innovative activity complicates the picture. A large number of American firms that have exhibited high propensities to innovate have been recipients of such funding at one level or another. This has been true of large firms such as the large aerospace companies and smaller firms such as those of Route 128 and the Santa Clara Valley. The funding has not necessarily always been direct; often, especially in the case of smaller firms, it has been channeled by means of subcontracting.

The effects of governmental intervention in the development of technology have not been well studied, and opinions on the effects are not unanimous. At one extreme, J. J. Servan-Schreiber, in his best-selling book, *The American Challenge*, identified federal support of industrial R&D as a major source, perhaps *the* major source, of what he perceived as U.S. economic dominance of Europe.[32] Servan-Schreiber did not, however, present much in the way of empirical evidence to support this contention. Arguments similar to those of Servan-Schreiber have been advanced by more scholarly observers.[33]

By contrast, J. Herbert Hollomon argues that during the 1950's and 1960's, the federal government supported R&D in certain areas of high technology, most notably technologies associated with the space and atomic energy programs and with the development of strategic weaponry.[34] The support was so intense, Hollomon observes, that an excessive demand was placed on specialists in high technology. This had two adverse long-term effects on the economy. First, the supply of qualified persons trained in commercial technologies (those of such basic industries as metallurgy and machinery) was suppressed, and the factor cost of such personnel accordingly rose. This, Hollomon maintains, had the effect of suppressing technological innovation in many industries in which R&D was not funded by governmental programs. Second, the U.S. technical education system became attuned to the "production" of individuals trained in the aerospace technologies, and when demand for such individuals slackened in the late 1960's and early 1970's, the educational system was unable to adjust.

Unfortunately, there does not exist much empirical evidence shedding light on this controversy. Both arguments have some ring of truth. On the one hand, U.S. governmental support of R&D during the years following the Korean War but prior to the Vietnam War was unquestionably strongly oriented toward the development of the so-called high technologies applicable to defense, space, and atomic energy programs. This may have distorted allocation of R&D resources. On the other hand, the "fallout" of commercially useful innovation from the aerospace and related programs has been of significance in some industries, most notably air transport and electronics. It is doubtful that anyone can say at this point in time whether the total benefits justify the total costs and distortions.

Robert Gilpin, in a widely read report to the Joint Economic Committee of the U.S. Congress, has suggested that there are three valid justifications for governmental funding of R&D and two nonvalid (but frequently articulated) ones.[35] The justifiable reasons, in Gilpin's view, are as follows: (1) "The Public Nature of Knowledge." If the expected social returns to an innovation are positive and large enough but the private returns are too small to stimulate entrepreneurs to undertake the innovation, the government should subsidize the innovation. (2) "Structural Aspects of Industry." If established patterns of firm conduct in an oligopolistic industry are such that innovation is discouraged, the government might take action to stimulate innovation through subsidization. Likewise, if the average size of firms in an industry is small and economies of scale in R&D are large, the government might undertake the role of the innovator. This occurs in sectors of agriculture, for example. (3) "Social and Political Needs." If there are needs for innovation that do not even generate a positive social rate of return, the government must undertake to fulfill those needs.

Economists would quarrel with Gilpin's third justification. If social returns were calculated on the basis of correctly determined shadow prices, there would never be any basis for undertaking a project that generates a negative social rate of return. Such a project would at best be a "boondoggle."

In Gilpin's view, the nonjustifiable reasons are as follows: (1) "The Scale Argument." It is frequently argued that certain projects must be government funded because the scale of the projects is too great to permit private enterprise to bear the risks of the development. Gilpin argues that unless the social returns are positive and large enough, the project should not be undertaken.
(2) "The Security of Supply." Gilpin argues that funding of technologies to develop substitutes for resources for which the nation is dependent on foreign sources of supply

is unjustifiable. The argument is frequently made that for strategic reasons the United States must develop substitutes for imported resources. Gilpin argues that in most cases the costs of development would far outweigh the strategic benefits that would accrue from governmentally subsidized programs to make the United States "self sufficient" in these resources.

Funding of R&D is not, of course, the only means by which the government can affect technological change. Governmental regulatory policy can have a profound effect on the rate and direction of innovative activity, and this effect is not always necessarily an intended one.

Somewhat surprisingly, perhaps, the number of empirical studies of the effects of governmental regulation upon technological innovation is not large. A number of studies in the pharmaceutical industry report the negative impact of stringent FDA regulations upon the amount of pharmaceutical research being conducted in this nation.[36] The suggestion is that federal regulation in this industry has slowed the development of new drugs. Typical of complaints voiced by the pharmaceutical industry are that the complexity and specificity of FDA procedural requirements for new drug applications have increased dramatically in recent years and that the time and cost of complying with these requirements has become excessive. One study notes that the amount of documentation that must accompany a new drug application increased from a typical 500 pages in 1958 to many tens of thousands of pages in the early 1970's.[37] It has also been noted that the average time of development of a new drug increased from about 2 years in the 1958-1962 period to about 8 years in the 1970's.

While a slowdown in the rate of new drug introduction is a documentable fact, it is not totally clear that FDA regulation is wholly responsible for this slowdown. While there is evidence that the total rate of successful innovations has declined from the heyday of new drug development of the 1950's and 1960's, some industry spokesmen attribute this to the fact that new drugs are simply becoming much harder to develop as well as to the effects of regulation.[38]

There is little doubt that governmental regulation makes life difficult for entrepreneurs in a number of industries.[39] It can be the case, however, that "making life difficult" can lead to increased incentives for technological innovation rather than the reverse. Regulations designed to protect the natural environment are a case in point. Strict enforcement of antipollution and other environmental regulations, without question, increases the need for the development of new technologies and hence can lead to increases in the rate of innovation.[40] Whether or not the benefits of the investment in this innovation justify the total cost is a debatable issue that is discussed further in later chapters.

5. TECHNOLOGICAL INNOVATION AND PRODUCTIVITY

Much of the workshop discussion centered upon the effect of technological innovation upon U.S. productivity. It was recognized that advances in productivity (i.e., increases in output per unit of factor input) in the long run are determined by advances in technology. In the short run, however, productivity might be affected by a host of variables other than technological ones. This section reviews some of these.

Various forces are generally believed to lie behind changes in productivity. One force is changes in the mix and quality of factor inputs into the economy. By aggregating the factors into three broad categories of natural resources, labor, and capital, the nature of these changes can be explored.

One possible cause of changing rates of productivity is a change in the quality of natural resource factor inputs. For many natural resources, especially those that fall into the category of mineral, it is asserted that the availability of easily accessible, rich grades of resources is declining and that, as a consequence, the world's economies are forced to exploit lower grades or less accessible resources. Taken *in isolo*, the forced substitution of lower grades of natural resources for higher-grade ones would lead to reduced productivity. Until the late 1960's, reductions in the quality of natural resources were more than offset by advances in the technology of resource extraction. The limited evidence available, however, suggests that such offsets might be harder to come by during the 1970's.

Changes in the composition of the labor factor certainly can affect productivity. The case has been made by several analysts that the much celebrated, post-1966, U.S. "productivity slowdown" has been a consequence in part of the addition of women, youths, and minorities to the U.S. work force at an advancing rate.[41] These workers, it is asserted, are often not as skilled or experienced as older male workers and hence do not produce as much output per hour worked. Offsetting this is the fact that these new additions to the labor force are often compensated at lower rates than are older workers. If lower wages of the youth, women, and minorities are a result of less experience and hence lower output per hour worked, little or no downward bias in reported productivity per unit cost of labor will result. If, on the other hand, the lower wages are in effect discriminatory, reported productivity in those industries in which output is measured in terms of cost of input might in fact be less than real productivity.

In labor-intensive service industries, for example, the reported value of output is heavily affected by wage rates. Thus, in these industries, measured productivity is largely a function of the wage rate. If there are barriers to entry to the labor force in an industry, so that wages can include a rent to the worker, a rise in the wage rate might be accomplished without a rise in the quantity or quality of services performed per hour worked. Thus, in such an industry, reported productivity might increase faster than real productivity.

It should be noted that, overall, the quality of the U.S. labor force, at least as measured by means of productivity statistics, has been steadily on the increase since World War II. The increase is commonly believed to be the consequence of better and more extensive education and improved health standards. In the forthcoming decade, it is believed that any recent, short-run reduction in the rate of measured improvement in the quality of the U.S. labor force will likely reverse itself as (1) the age-sex-education characteristics of the labor force stabilize, (2) recent new entrants to the labor force (especially women, youths, and minorities) acquire the skills necessary to bring their output per hour worked up to levels equivalent to those of the established members of the work force, and (3) wage discrepancies between the sexes and among ethnic groups disappear.

Just as the quality of the U.S. work force has steadily improved since World War II, the quality of the work forces of Germany, several other European nations, and Japan has improved even more dramatically during the same time. The fact is that levels of health and education in these nations lagged well behind those of the United States in 1950 but by 1977 the situation had changed dramatically. The rapid improvement in the quality of the labor force in nations such as Germany and Japan doubtlessly has been one major factor behind the rapid increases in productivity recorded in these nations.

The quality of the stock of capital goods of a nation

also can affect productivity. The quality of capital goods is a function of both their age and their design. Presumably, newer goods tend to embody more recent technology than do older goods, resulting in the newer goods being more efficient. Additionally, newer goods on balance ought to be subject to less downtime per hour of operation than are older goods. Thus, because of embodied technology and maximum attainable utilization rates, new capital goods ought to be more efficient than are older ones.

It is through improvements in the quality of capital that technological innovation most profoundly affects productivity, because new process technology enables greater unit output to be achieved per unit of factor input. However, realization of the benefits of technological innovation in an economy is not instantaneous. The rate of realization depends upon the rate of diffusion of the innovation. It is only when new capital is added to the economy, or when old capital is replaced by new capital, that the potential increases in productivity are realized.

A related but separable matter is that of economies of scale. Scale economies imply that at high levels of output, factor usage per unit of output is lower than at lower levels of output. Thus, if all other things are equal, an economy having capital stock that embodies economies of scale will exhibit a higher productivity than one whose capital stock does not embody economies of scale. The significance of this can be illustrated by the following hypothetical example: suppose that there exist two nations, each having qualitatively identical populations, equal factor proportions, and equal levels of technological knowledge, but that one nation is much larger than the other. If the larger of these nations is able to achieve higher levels of scale economies than is the smaller nation, the larger nation will exhibit a greater productivity than will the smaller, in spite of all other conditions being identical. In this case, the larger nation would in a sense have a higher "quality" of capital stock, even if this capital stock embodied no technology that was not available to the smaller nation. It is thus important to note that economies of scale do not necessarily imply a more advanced level of technology. Increased potential for economies of scale can, however, result from technological innovation.

Changes in the quality of factor inputs are not the only source of changes in productivity. Changes in output mix are a source of productivity change that might have an important bearing on total U.S. productivity. It has been noted by many analysts that the total mix of U.S. final output has been shifting steadily away from the manufacturing industries and toward the service industries for at least two and a half decades. Because the value of the service industries' output is usually measured in terms of the value of factor input, these industries typically report the lowest productivity advances of all industries. Thus, the shift in the U.S. economy from manufacturing industries to service industries has an adverse impact upon reported rates of U.S. productivity growth.

Intangible factors doubtlessly affect productivity growth. Attitudes of a nation's population and national work habits probably play a major but largely unmeasurable role.[42] A lackadaisical attitude toward work is often cited in the press as a major force behind Great Britain's recent economic performance, and (in the U.S. press) the worry is continually being raised that the United States may be experiencing similar problems. Japan's enormous economic growth is often popularly ascribed to the enthusiastic work habits of Japanese workers. There may be anywhere from some to a lot of truth in such assertions, but, given the present state of the art of relating social factors to economic performance, the assertions must be regarded as somewhat speculative.

Short-term forces can affect reported rates of productivity. Prime among these short-term forces would be changes in rates of capacity utilization of individual plants or industries. If the change in the capacity utilization were to converge upon the most efficient rate, reported productivity would clearly increase, while if the change were to be away from the most efficient rate, the reported productivity would decrease. Changes in rate of capacity utilization are typically cyclical in nature, and hence, to be meaningful, productivity data should be cyclically adjusted.

Other short-run forces can be noted. The learning curve phenomenon might affect the short-run productivity of a new technology. As the new technology is introduced to a plant or an industry, the technology might not be efficiently utilized until the workers have learned how to cope with the changes brought on by the new technology. Cyclicality in the rate of investment in new plant and equipment might have a short-run effect upon the rate of diffusion of new technology. During the peak of the capital investment cycle, new plants and equipment embodying new technology will be added to the capital stock at a more rapid rate than during the trough of the cycle.

6. TECHNOLOGY AND INTERNATIONAL TRADE

Historically, the branch of economic theory devoted to the study of international trade has largely operated under the assumption that the technological knowledge available to all nations engaged in international trade is equal. As implausible as this assumption is, much of the theory of international trade is premised upon such an assumption. Under this assumption, trade between nations occurs as a result either of differing consumer tastes or of differing endowments of the tangible factors of production (i.e., land, labor, and capital).[43]

It is little short of obvious, however, that the technological capacities of the earth's nations are very disparate and that the international diffusion of technology occurs at a less than instantaneous rate. The technology factor in international trade was first approached by economic theorists during the 1950's. The central concern was how technologically induced changes in factor endowments would affect a trading nation's terms of trade.[44] In a two-nation, two-good, two-factor trading model, the effects would be as follows:[45]

Assume that the two nations initially possess the same technology and that technological change occurs in the nation in which capital is relatively abundant. The technological change is assumed to be costless and to increase the efficiency of manufacture of one or both of the goods. The demand for both goods is characterized by non-negative income elasticities. If the technological change is nonfactor biased (factor neutral), and if it occurs in the capital-intensive good, it will result in "ultra-export-bias," that is, both the level of production and the level of exportation of the capital-intensive good will increase and both the level of production and the level of importation of the labor-intensive good will decline. Consequently, the terms of trade will decline. If the technological change is nonfactor biased and it occurs in the labor-intensive good, it will result in "ultra-import-bias," and the terms of trade will improve. If the technological change is capital using in the labor-intensive good, ultra-import-bias will again occur, and if the technological change is labor using in the capital-intensive good, ultra-export-bias will again occur. If the technological change is labor using in the labor-intensive good or capital using in the capital-intensive good, the effects on the terms of trade cannot be ascertained without consideration of the demand characteristics of the products.

Later theoretical work expanded this type of analysis to cases wherein technological change is not considered to be costless and to cases wherein there are more than two goods and two factors.[46]

Empirical evidence that differing levels of technological knowledge among nations might be an important determinant of patterns of international trade is, in fact, quite recent, the major articles having appeared in the international economics journals only in the 1960's.[47] Common to many of these articles was the observation that some large proportion of exported manufactured goods of the United States was characterized by a large "R&D factor," that is, that amortized research and development expense constituted a large portion of the added value of these products. That U.S. exports might to a large extent be R&D intensive was presented as one possible explanation to the so-called Leontief paradox, the empirical evidence that U.S. exports were on the balance more labor intensive than domestically produced manufactured goods that were also imported.[48] (Neoclassical trade theory would predict the opposite, that a capital-rich nation such as the United States would export goods that were relatively capital intensive and import goods that were relatively labor intensive.)

One of the strongest advocates of the idea of the technology factor in international trade was Raymond Vernon, who advanced one set of hypotheses to explain the reason why U.S. exports in manufactured goods tend to be R&D intensive.[49] This set of hypotheses, generally known as the "product cycle" theory of international trade, is structured as follows. An extension of Schmookler's hypotheses, that a large and growing demand for a product stimulates technological innovation in the design and manufacturing of the product, is assumed to be valid, and it is further assumed that innovation in process technology will generally result in new processes that use the abundant factors of production.[50] Furthermore, it is assumed that innovation in product technology will produce new consumer products that, initially at least, will be highly income elastic in their demand characteristics. Thus, these new products will find their initial markets to be concentrated among high-income consumers.

Proponents of the product cycle hypothesis maintain that it is a particularly valid device to explain U.S. export performance with respect to Western Europe during the 1950's and 1960's. During this time, the United States was by far both the largest market and the market with the highest per capita income in the world. The U.S. economy, in relation to the rest of the world, was also capital intensive and thus relatively scarce in labor. It is argued by Vernon that labor scarcity was particularly the case for specialized, artisan-class workers.[51]

It is asserted that these characteristics of the U.S. market were at this time unique among the world's nations. The expectation therefore was that the United States would develop innovations characteristically unique to the United States, ones that simply would not be developed elsewhere. In other nations, however, there might be a demand for these unique products and processes. The demand would arise from two sources: (1) high-income consumers, located in overseas markets of insufficiently large internal size to stimulate the innovation of income-elastic consumer products and (2) producers seeking labor-saving technologies. These producers would typically be located in nations that, like the United States, were relatively underendowed with labor.

In the nations of Western Europe in particular, where per capita income was growing at a rapid rate during the 1950's and 1960's and where factor endowments were shifting toward increasing capital-to-labor ratios, the demand for U.S. product innovations would be large and growing. At the outset, at least, the demand was satisfied by exportation. However, over time, as the economic characteristics of the European market *and* the economic characteristics of these products change, trade patterns will change as well.

The economic characteristics of a product of innovation changes as the design of the product becomes standardized. Three consequences are asserted: (1) the product can be manufactured by standardized processes for which there are increasing returns to scale, (2) these processes tend to be capital intensive, and (3) as time passes, the product is increasingly easily imitated. Because of this latter consequence, the likelihood increases with time that the product (or close substitutes) can be produced by firms other than the innovating firm.

The net result is that the product can be produced at an increasingly lower cost, and the price (at a given level of demand) falls. New entry into the industry reduces rents and results in further price reductions. (In effect, over time, the supply curve for the product shifts to the right.) If demand for the product is price elastic, output will expand, the growth rate being a function of the rate of shift of the supply curve and the price elasticity of demand for the product. As the product matures, however, opportunities for cost reduction through standardization and increasing returns to scale will diminish, and hence the supply curve will stabilize.[52] Thus, over time, it is asserted, the economic characteristics of the product change in a qualitatively predictable fashion.

In Western European markets, growth in per capita income and rising capital-to-labor ratios, it has been asserted, created a growing demand for products of U.S. innovation. Thus, changes in the economic characteristics of the European market created a growing demand in Europe, which was coupled with a growing supply and declining barriers to entry for these products. The result, according to the hypothesis, is that new entry into the manufacture of the products occurs in Europe. The new entrants may be local firms, or they may be local subsidiaries of the original exporting American firms set up to "defend" the European market from local new entrants. In either case, the net effect is that, over time, U.S. exports of any given innovated good to Western Europe decline. As the good becomes economically mature, the basis for comparative advantage increasingly ceases to be technological innovation (the basis for the initial comparative advantage of the United States) and becomes relative factor cost. Thus, over time, comparative advantage changes. If the United States has a long-run comparative advantage in the manufacture of a given product, then, as the product matures, the United States will continue to export the product to Europe even though the product might also be manufactured in Europe. If long-run comparative advantage for a particular product favors Europe, U.S. exports to Europe of this product eventually will cease altogether and the trade patterns will reverse themselves: Europe will export to the United States. Even so, the United States may continue to manufacture the product domestically.[53]

The product cycle hypothesis was advanced primarily to explain certain empirical patterns in U.S. exports of manufactured goods, and empirical tests of the hypothesis have generally corroborated it.[54] There is no reason to expect the United States to be the sole producer of technologies embodied in exportable goods, however, and it has been suggested that the product cycle hypothesis might be capable of explaining exports of other nations in which conditions exist for the occurrence of large amounts of technological innovation.

Western Europe and Japan have certainly both passed the point in the post-World War II era of being largely users or imitators of U.S. technological innovation, and both regions are now fully capable of undertaking their own innovative efforts.[55] The implications of the rise (or re-emergence, to be more accurate) of these regions as technological innovators in their own right are discussed in the following chapter.

The product cycle is advanced as a working hypothesis of how the technology factor affects international trade. Proponents of the hypothesis do not claim that technology is the only factor affecting trade. For highly standardized exportable goods, the neoclassical (Heckscher-Ohlin) hypothesis of international trade is quite powerful.[56] The United States, rich in arable land and in agricultural capital, does indeed export large quantities of agricultural commodities produced by capital-intensive processes, and Colombia, a nation also well endowed with land but less well endowed with capital, exports relatively labor-intensive agricultural products. As was suggested in the previous paragraph, the basis for comparative advantage in a manufactured good, however, might change over time, so that a nation that initially is a net importer or exporter of the good eventually becomes a net exporter or importer.

NOTES

1. The literature on factor bias in technological innovation is quite large. For a bibliography, see John S. Chipman, "Induced Technical Change and Patterns of International Trade," in Raymond Vernon, editor, *The Technology Factor in International Trade* (New York: National Bureau of Economic Research, 1970). The classical works are J. R. Hicks, *The Theory of Wages* (New York: Macmillan Publishing Company, Inc., 1935) and J. Robinson, "The Classification of Inventions," *Review of Economic Studies* 5 (1937-1938). An excellent summary of the concepts is to be found in R. Findlay and H. Grubert, "Factor Intensities, Technological Progress, and the Terms of Trade," *Oxford Economic Papers*, 1959, pp. 111-121, reprinted in Jagdish Bhagwati, editor, *International Trade: Selected Readings* (New York: Penguin Books, Inc., 1967).

2. Whether or not U.S. technological innovation has been capital using is a matter of some debate. For contrasting views, see Charles Kennedy, "Induced Bias in Innovation and the Theory of Distribution," *Economic Journal*, September 1964, and W. E. G. Salter, *Productivity and Technical Change* (New York: Cambridge University Press, 1964). For empirical studies addressing this issue, see J. L. Enos, "Invention and Innovation in the Petroleum Refining Industry,"

in R. R. Nelson, editor, *The Rate and Direction of Inventive Activity* (Princeton: Princeton University Press, 1962); M. J. Piore, "The Impact of the Labor Market Upon the Design and Selection of Productive Techniques Within the Manufacturing Plant," *Quarterly Journal of Economics*, November 1968; and William H. Davidson, "Patterns of Factor Saving Innovation in the Industrialized World," *European Economic Review*, December 1976.

3. It should be noted that although this implies that wages will fall relative to returns to capital, the absolute level of wages in real terms may actually rise owing to greater factor productivity.

4. This line of reasoning is pursued in G. Bitras, K. Lee, and F. Machlup, "Effects of Innovations on the Demand for and Earnings of Productive Factors," Mimeo (Washington, D.C.: National Science Foundation, 1976).

5. See C. Kennedy, "Induced Bias in Innovation and the Theory of Distribution," *Economic Journal*, September 1964.

6. See *Surveys of Business' Plans for R&D Expenditures* (New York: McGraw-Hill, issued periodically).

7. Many product innovations lead to increases in efficiency through substitution. Thus, for example, the introduction of synthetic fibers into the textile industry enabled the production of fabrics that were lower in cost and lighter in weight than those previously possible and hence to an increase in economic efficiency in this industry. As Simon Kuznets points out, however, to treat the development of synthetic fibers solely as an increase in efficiency of the textile industry is to overlook qualitative economic gains resulting from this development. See Simon Kuznets, "Inventive Activity's Problems of Definition and Measurement," in R. R. Nelson, editor, *The Rate and Direction of Inventive Activity* (Princeton: Princeton University Press, 1962). See also W. Eric Gustafson, "R&D, New Products, and Productivity Change," *American Economic Review Papers and Proceedings*, May 1962, and Zvi Griliches, "Comment," *ibid.*, for a discussion of this issue.

8. See Joseph A. Schumpeter, *History of Economic Analysis* (New York: Oxford University Press, 1954) and "The Analysis of Economic Change" in *Readings in Business Cycle Theory* (London: Blakiston, 1944).

9. The notion that technological invention was an activity brought about by man's insatiable curiosity and drive to create new devices was commonly accepted in the 1930's, and hence Schumpeter's assumption was not without basis. However, as early as 1870, John Stuart Mill had suggested that inventive activity might be induced by the prospect of

economic gain as well as by pure curiosity and thus that inventive activity could not be uncoupled from economic activity. See J. S. Mill, *Principles of Political Economy* (New York: D. Appleton Publishing Company, 1890).

10. See Edwin Mansfield, *Technological Change--An Introduction to a Vital Area of Modern Economics* (New York: Norton Publishing Company, 1971), and Nathan Rosenberg, "Factors Affecting the Payoff to Technological Innovation," Mimeo (Washington, D.C.: National Science Foundation, 1976), pp. 13-29. But also see Kenneth J. Arrow, "Comment," in Raymond Vernon, editor, *The Technology Factor in International Trade* (New York: National Bureau of Economic Research, 1970).

11. For empirical evidence, see W. F. Mueller, "The Origins of the Basic Inventions Underlying DuPont's Major Product and Process Innovations, 1920 to 1950," in R. R. Nelson, editor, *The Rate and Direction of Inventive Activity* (Princeton: Princeton University Press, 1962).

12. See L. Nasbeth and G. F. Ray, *The Diffusion of New Industrial Processes* (New York: Cambridge University Press, 1974) for a series of case studies of the diffusion of new process technology.

13. J. L. Enos, "Invention and Innovation in the Petroleum Refining Industry," in R. R. Nelson, editor, *The Rate and Direction of Inventive Activity* (Princeton: Princeton University Press, 1962).

14. Rosenberg, *Technological Innovation*.

15. Jacob Schmookler, *Invention and Economic Growth* (Cambridge, Mass.: Harvard University Press, 1966).

16. *Ibid.*, p. 10.

17. For a survey, see M. I. Nadiri, "Some Approaches to the Theory and Measurement of Total Factor Productivity: A Survey," *Journal of Economic Literature*, December 1970. See also Dale Jorgenson and Zvi Griliches, "The Explanation of Productivity Change," *The Review of Economic Studies*, July 1967, and a comment on this article by E. F. Denison in *Survey of Current Business*, May 1969.

18. The estimation procedure involves the use of a production function (usually a Cobb-Douglas function) for which parameters are estimated for the U.S. economy. Time series data for the U.S. capital stock and labor supply are inputed into the production function, and an estimated output is calculated. The difference between the estimated output and the actual output of the economy is the residual.

Credit for the first sophisticated effort to measure the effect of technological change on the U.S. economy is generally given to Robert Solow. Since Solow's efforts, numerous variants on Solow's methodology have been tried. See R. M. Solow, "Investment and Technical Progress," in K. J. Arrow, S. Karlin, and P. Suppes, editors, *Mathematical Methods in the Social Sciences* (Stanford, Calif.: Stanford University Press, 1969), and "Technical Progress, Capital Formation, and Economic Growth," *American Economic Association Papers*, May 1962.

19. See Evsey Domar, "On the Measurement of Technological Change," *Economic Journal*, December 1961.

20. E. F. Denison, *Why Growth Rates Differ* (Washington, D.C.: Brookings Institution, 1967).

21. See Dale Jorgenson and Zvi Griliches, "The Explanation of Productivity Change." *The Review of Economic Studies*, July 1967.

22. See F. M. Scherer, *Industrial Market Structure and Economic Performance* (Chicago, Ill.: Rand McNally and Company, 1970), pp. 346-347.

23. See E. Mansfield, J. Rapaport, A. Romeo, S. Wagner, and G. Beardsley, "Social and Private Rates of Return from Industrial Innovations." *Quarterly Journal of Economics* 2 (May 1977).

24. F. M. Scherer, "Firm Size and Patented Inventions," *American Economic Review*, December 1965.

25. F. M. Scherer, *Industrial Market Structure*, chap. 15.

26. J. Jewkes, D. Sawers, and R. Stillerman, *The Sources of Invention* (New York: Macmillan Publishing Company, Inc., 1969).

27. For a somewhat heroic portrayal of several such entrepreneurs, see Gene Bylinsky, *The Innovation Millionaires* (New York: Charles Scribner and Sons, 1976).

28. W. R. Maclaurin, *Invention and Innovation in the Radio Industry* (New York: Macmillan Publishing Company, Inc., 1949) and S. G. Sturmey, *The Economic Development of Radio* (London: Duckworth, 1958).

29. For an example, see M. J. Peck, "Inventions in the Postwar American Aluminum Industry," in R. R. Nelson, *The Rate and Direction of Inventive Activity* (Princeton: Princeton University Press, 1962).

30. Such would apparently be the case in several industries, including electronic computers, microwave communications, and household chemicals. See "I.B.M.'s $5,000,000,000 Gamble," *Fortune*, September 1966; F. M. Scherer, "The Development of the TD-X and TD-2 Microwave Radio Relay Systems in Bell Telephone Laboratories," Mimeo Harvard Business School Case Study (Cambridge, Mass.: Intercollegiate Case Clearinghouse, 1960); "Lestoil: The Road Back," *Business Week*, June 15, 1963.

31. Such was the conclusion of S. G. Sturmey in his study

of the radio industry. Sturmey noted that new entrants into the industry rarely brought significant new technology with them, but that their entry induced the industry leaders to innovate. See S. G. Sturmey, *Economic Development of Radio*.
32. See J. J. Servan-Schreiber, *The American Challenge* (New York: Atheneum, 1968; translated from the French book entitled *Le Défi Américain*, 1967).
33. The point is made in Raymond Vernon, *Sovereignty at Bay* (New York: Basic Books, 1970), chap. 3.
34. J. Herbert Hollomon, "America's Technological Dilemma," *Technology Review*, July-August 1971.
35. Robert Gilpin, *Technology, Economic Growth, and International Competitiveness*, A Report to the Joint Economic Committee of the U.S. Congress (Washington, D.C.: U.S. Government Printing Office, 1975), chap. V.
36. See Lewis A. Sarett, "FDA Regulations and their Influence on Future R&D," *Research Management*, March 1974, and Charles C. Edwards, "The Role of Government and F.D.A. Regulations in Drug R&D," *Research Management*, March 1974, for conflicting views on this subject. See also Mary Heston-Sands and Lawrence L. Hope, "Strategy and Planning in a Turbulent Environment: The Ethical Pharmaceutical Industry" (unpublished S.M. Thesis, MIT Sloan School of Management, Cambridge, Mass., 1976) and bibliography therein.
37. See Joseph F. Sadusk, Jr., "The Effect of Drug Regulation on the Development of New Drugs," in F. Gilbert McMahon, editor, *Principles and Techniques of Human Research and Therapeutics* (Mount Kisco, N.Y.: Futura Publishing Company, 1974) vol. 1.
38. National Research Council, *Technology Transfer From Foreign Direct Investment in the United States*, Report of a Seminar Series (Washington, D.C.: National Academy of Sciences, 1976), chap. 1.
39. See the "Survey of Governmental Regulation" reported in *Business Week*, April 4, 1977.
40. See Stanley M. Greenfield, "Incentives and Disincentives of EPA Regulations," *Research Management*, March 1974. An unpublished study by the MIT Center for Policy Alternatives suggests that the net benefit to the economy from the Environmental Protection Agency regulation has been strongly positive and that the benefits include increased efficiency at the plant level as well as cleaner air and water.
    Commenting upon the effect of governmental regulation in the telecommunications industries, Dean Gillette has noted that the overall impact has been to stimulate useful innovation. However, Gillette cites cases of overzealous regulation that have stifled innovation. See Dean Gillette,

"Innovation Under Regulation," paper presented at panel on "The Effect of Government Antitrust Action and Regulation on Technological Innovation: The Issues," American Association for the Advancement of Science, Annual Meeting, Washington, D.C., February 20, 1976.

41. See, for example, Edward F. Denison, *Accounting for United States Economic Growth* (Washington, D.C.: Brookings Institution, 1974).

42. See John W. Kendrick, "Productivity Trends and Prospects," paper prepared for Joint Economic Committee of the United States Congress, Washington, D.C., June 1976.

43. See, for example, R. Caves and R. Jones, *World Trade and Payments* (Boston: Little, Brown and Company, 1973).

44. The early works include the following: J. R. Hicks, "An Inaugural Lecture," *Oxford Economic Papers*, vol. 5, 1953; William M. Corden, "Economic Expansion and International Trade: A Geometric Approach," *Oxford Economic Papers*, vol. 8, 1956; and Harry G. Johnson, "Economic Expansion and International Trade," *Manchester School of Economic and Social Studies*, May 1955.

45. These are derived in R. Findlay and H. Grubert, "Factor Intensities, Technological Progress, and the Terms of Trade," *Oxford Economic Papers*, vol. 11, 1959.

46. See P. K. Bhardan, "On Factor-Biased Technical Progress and International Trade," *The Journal of Political Economy*, August 1965, and John S. Chipman, "Induced Technical Change and Patterns of International Trade."

47. See, for example, W. Gruber, D. Mehta, and R. Vernon, "The R&D Factor in International Trade and International Investment of United States Industries," *The Journal of Political Economy*, February 1967, and D. B. Keesing, "The Impact of Research and Development on United States Trade," *The Journal of Political Economy*, February 1967.

48. See Mordechai Kreiniz, "The Leontief Scarce-Factor Paradox," *American Economic Review*, March 1965, for an exposition of Leontief's findings and comments thereon.

49. See Raymond Vernon, "International Investment and International Trade in the Product Cycle," *Quarterly Journal of Economics*, May 1966.

50. For evidence see W. H. Davidson, "Patterns of Factor Saving Innovation in the Industrialized World," *European Economic Review*, December 1976.

51. See Raymond Vernon, *Sovereignty at Bay*, chap. 3.

52. This assumes that demand characteristics of the product remain stable. If demand conditions change, these may induce further changes in supply.

53. The implications of this are discussed by A. J.

Karchere, "The Effect of Transnational Companies on the U.S. Economy, and Future Prospects," in *The International Essays for Business Decision Makers* (Dallas: Southern Methodist University, 1976). Karchere emphasizes the export role of the United States in products in which the United States holds a comparative advantage.

54. See Louis T. Wells, "Test of the Product Cycle Model of International Trade," *Quarterly Journal of Economics*, February 1969; Louis T. Wells, editor, *The Product Life Cycle and International Trade* (Cambridge, Mass.: Harvard University Press, 1973); and Robert B. Stobaugh, "The Product Life Cycle, U.S. Exports, and International Investment" (unpublished Ph.D. thesis, Harvard University, Cambridge, Mass., 1968).

55. For Europe, see Larry Franko, *The European Multinationals* (New York: Harper and Row, 1976), as well as references in the following chapter. For Japan, see Michal Y. Yoshino, *Japan's Multinational Enterprises* (Cambridge, Mass.: Harvard University Press, 1976), and Yoshi Tsurumi, *The Japanese are Coming* (Cambridge, Mass.: Ballinger Publishing Company, 1976).

56. According to Harry G. Johnson, both the Vernon hypotheses and the Heckscher-Ohlin hypotheses are valid and each captures only part of the truth in international trade. See Johnson, "Technological Change and Comparative Advantage: An Advanced Country's Viewpoint," *Journal of World Trade Law*, January-February 1975.

3 TECHNOLOGY TRANSFER AND TRADE BETWEEN THE UNITED STATES AND THE OTHER OECD NATIONS: CRITICAL ISSUES

Among the central topics of the workshop were two interlocking questions. First, has the United States lost, or is it in the course of losing, its "lead" in the innovation of new commercially applicable technology? Second, why has productivity in the United States--both total factor productivity and real product per man-hour--been growing more slowly than in the majority of the other OECD nations?

The analysis of these two questions entailed an exploration of three subsidiary questions. First, to what extent is the slow rate of growth of productivity in the United States attributable to a diminution of the rate of technological innovation in this country? Second, should a reduced "lead" in the rate of technological innovation in the United States in relation to other OECD nations necessarily be a cause for alarm? And third, can any reduced lead by the United States in technological innovation be directly attributed to technology transfer from the United States to other OECD nations, either by multinational corporations or by other means?

On these questions there emerged a considerable range of opinions among the workshop participants. These opinions corresponded in substantial measure with differences in view on the same questions that have been aired in the current literature. Some of these varied views are presented in Appendix A.

On these questions, the participants tended to cluster about three somewhat different points of view. The first of these points of view was that the lead of the United States in the development of new commercial technology is diminishing, the loss is measurable, and the loss can in part at least be directly attributed to the transfer abroad of U.S. technology. Proponents of this view tended to believe that the loss by the United States of its technological lead is highly inimical to U.S. long-term interests.

A second point of view was that the United States is losing the lead but that the reasons for this have little to do with international technology transfer. Rather, the cause was seen to lie in a loss of dynamic, innovative vitality within the domestic U.S. economy, which is reflected in, among other things, reduced rates of improvement in U.S. productivity, lessened ability of U.S. industry to compete in international markets, and decreased willingness of the U.S. entrepreneurs to engage in risky ventures.

The third point of view was that the United States is losing the technological lead only in the sense that a small number of other nations have come to develop the internal capabilities to generate technological innovation on the large scale that the United States is capable of. According to this posture, the primary cause of the diminution of the U.S. lead is the fulfillment of European and Japanese capabilities, which did not come to full fruition until the 1970's because of the aftermath of World War II. Proponents of this third point of view tended to be less worried about the implications of a loss of lead by the United States for U.S. long-term interests than were proponents of the other two views. Nonetheless, even within the third group there was some concern that the dynamic vitality of the U.S. economy, perceived by virtually all discussants as the underlying force leading to technological innovation, might be faltering somewhat.

It was felt by the workshop participants that the existing body of knowledge is inadequate to make a conclusive case for any of these points of view. All that can be offered are partial analyses of a very complex problem. For example, although there is widespread opinion that the rate of technological innovation in the United States has faltered somewhat in recent years, there simply is no means to ascertain whether or not the United States is experiencing a long-term decline in its ability to innovate. The effects of technological transfer on the economies of both the technology-receiving and the technology-donating nations have been the subject of much research in recent years, but the total effects of technological transfer are highly complex and understanding of the process is far from complete.

Two concerns were shared by most participants in the workshop, irrespective of which point of view they held. The first was concern over the rate of increase of productivity in the manufacturing sector, a rate that has been consistently lower in recent decades in the United States than in most other OECD nations. The second was concern that the rate of new product introduction by established U.S. firms has waned in recent years and, furthermore, that the rate of formation of new enterprises created to manufacture and sell new products has come to a virtual standstill. Each of these problems is discussed in turn.

1. TECHNOLOGICAL INNOVATION AND U.S. PRODUCTIVITY (AND RELATED CONCERNS)

Much of the workshop discussion centered upon the effect of technological innovation upon U.S. productivity. Two aspects of productivity have captured the attention of U.S. policy makers in recent years. The first of these has been the so-called productivity slowdown following 1966. During the approximately two decades 1948-1966, total factor productivity grew at an average annual rate of some 2.5 percent, while real product per man-hour grew at an average annual rate of some 3.4 percent. During the years 1966-1973, the corresponding figures were approximately 1.6 percent and 2.3 percent, respectively. (See Table 1.) The second aspect of productivity is that the United States has, for at least two decades, consistently lagged behind most other OECD nations in terms of reported annual labor productivity increases. (Table 2 presents summary data on reported productivity increases of the major OECD nations.)

Discussion at the workshop focused upon how much of the post-1966 productivity slowdown and how much of the laggard productivity performance of the United States could be attributed to the nature and amount of U.S. technological innovation. It was recognized that to separate the effect of technological innovation upon productivity increases from the effects of other forces is at best a difficult matter and that different analysts who have attempted such a separation have not always reported consistent results. (See the discussion in Chapter 2.)

Two concerns expressed by the workshop were that the rate of innovative activity of the type that contributes to advances in productivity in the United States has been inadequate and that the rate of diffusion of productivity-advancing innovation has been slow. It was acknowledged that data that can be used to demonstrate either of these concerns definitively are not available, but the opinion of many of the workshop participants was that efforts to increase the rate of this innovation must be made.

Two hypotheses were advanced for the existence of a less than optimal rate of innovative activity. The first

TABLE 1 Productivity Trends in the U.S. Private Domestic Economy by Major Industry Divisions, Average Annual Percentage Rates of Change, 1948-1973, by Subperiods[a]

	Subperiod		
	1948-1966	1966-1969	1969-1973[b]
Private domestic economy			
Real product	4.0	3.4	3.8
Total factor productivity	2.5	1.1	2.1
Real product per unit of capital	0.4	-0.9	0.2
Real product per man-hour	3.4	1.7	2.9
Industry divisions (Real product per man-hour)			
Agriculture	5.6	6.7	5.3
Mining	4.6	1.8	0.2
Contract construction	2.0	0.0	-0.5
Manufacturing	2.9	2.7	4.5
Durable goods	2.8	2.2	--
Nondurable goods	3.2	3.4	--
Transportation	3.7	2.2	4.5
Communications	5.5	4.6	4.1
Electric and gas utilities	6.1	4.4	1.0
Trade	2.9	2.1	2.3
Wholesale	3.1	3.0	--
Retail	2.7	1.0	--
Finance, insurance, and real estate	2.1	-0.4	0.2
Services	1.2	0.4	1.0

[a]Subperiods are measured between successive business cycle peaks.
[b]Preliminary.
SOURCE: John W. Kendrick, *Postwar Productivity Trends in the United States* (New York: National Bureau of Economic Research, 1973). Estimates extended from 1969 through 1973 by the author.

TABLE 2 Productivity Growth of OECD Nations

Nation	Growth Rates of Gross Domestic Product per Person Employed		Growth Rates of Productivity by Sector, 1955-1968			Manufacturing Output per Man-Hour Worked	
	1960-1973	1969-1975[a]	Agriculture	Industry	Services	1960-1973	1973-1975
Canada	2.4	2.8	4.8	3.8	-0.1	4.3	-0.8
United States	2.1	2.6	5.4	2.9	1.7	3.3	-2.4
Japan	9.2	9.6	NA	NA	NA	10.5	6.2
France	5.2	5.2	6.1	5.3	3.4	6.0	3.6
Germany	5.4	4.4	6.1	5.0	2.5	5.8	4.0
Italy	5.7	5.3	7.8	5.8	3.7	6.4	NA
United Kingdom	2.8	3.1	5.8	2.9	1.4	4.0	0.1

NA = not applicable.
[a]Estimated.

SOURCE: *Expenditure Trends in OECD Countries, 1960-1980*, Table 5 (Paris: OECD, 1972); *The Growth of Output, 1960-1980*, Table 7 (Paris: OECD, 1970); "Report by OECD's Manpower and Social Affairs Committee," *OECD Observer*, March-April 1976, p. 10.

of these was that governmental regulation of various sorts has caused a net reallocation of innovative efforts away from those that might lead to productivity advances (or the availability of new products) and into efforts yielding nonproductive results. This matter is reported in the third section of this chapter.

The second hypothesis was that federal governmental intervention into the innovative process through allocation of resources into national defense and space programs contributes to a decline in productive advances in other sectors of the economy. This second hypothesis has been elaborated in the literature (see the discussion in Chapter 2), and little evidence exists either to corroborate or reject the hypothesis beyond that evidence already reported. Discussion of the hypothesis did lead, however, to a discussion of what might be the proper role of the U.S. government in the technological innovation process and, in particular, to how the federal government might stimulate innovation. This discussion is taken up in the third section of this chapter.

2. TECHNOLOGICAL INNOVATION IN NEW PRODUCT DEVELOPMENT

As was discussed in the first chapter of this report, technology can be loosely categorized as either product technology or process technology. Improvements in the latter, which is the knowledge associated with how to manufacture products, results in advances in productivity, the topic touched upon in the previous section. Improvements in the former can lead to productivity increases, especially if a newly created product is used as part of the process to manufacture some other product, but can also lead to qualitative improvements in the fulfillment of the needs of society.

Advances in productivity result in economic growth by reducing the total amount of resources required to make a given amount of any given end product, thus freeing resources and enabling them to be employed to other ends. Better fulfillment of the needs of society results in economic growth by utilizing resources in socially more desirable ways than was previously possible. Although these two sources of economic growth are not independent of one another to the extent that they are separable, the latter has probably contributed more to postwar economic growth in the United States than has the former. The spectacular growth of industrial sectors in the United States,

TABLE 3 Public Issues of Common Stock by New Small Companies

Year	All Small Companies		Small Companies Engaged in Technologically Intensive Activities	
	Number of Issues	Value of Issues[a]	Number of Issues	Value of Issues[a]
1969	649	1,103	204	349
1970	210	386	86	149
1971	244	528	73	138
1972	418	921	104	194
1973	67	158	19	38
1974	9	16	4	6
1975[b]	1	4	0	0

[a]Millions of dollars.
[b]First half of 1975.
SOURCE: J. O. Flender and Richard Morse, *The Role of New Technical Enterprises in the U.S. Economy* (Cambridge, Mass.: MIT Development Foundation, 1975).

such as the chemical, electronic, computer, plastic, and aerospace industries, has largely been a result of technological innovation resulting in new products that simply did not exist before.

Of concern to the workshop, therefore, was the possibility that the rate of this innovation might be on the decline. Two sets of evidence were introduced to suggest this possibility. The first was a series of public announcements by large U.S. corporations that R&D programs to develop new products are being slowed or curtailed. Such announcements have been made by firms that, historically, have been responsible for substantial achievements in the area of new product development. The other was inferential evidence that the rate of formation of new business firms engaged in producing technologically innovative products has declined drastically. (See Table 3.)

Numerous possible reasons were cited for this possible decline.[1] A listing of these would include the following:

1. Inflation. It was felt by most participants that monetary inflation has a detrimental effect on rates of

new product innovation. The detrimental effect of inflation stems from two sources. First, the expectation arises among businessmen that costs will rise faster than revenues, which causes businessmen to find ways to cut expenditures and to become increasingly averse to risk. Because expenditures for new product development have uncertain payouts, these expenditures are curtailed, often in favor of expenditures for development to cut the costs of the manufacture of existing products. Second, uncertainty grows with respect to the probable rate of increase in the cost of factor inputs required to produce new products. This is especially true for highly specialized inputs, such as custom-made machinery. Thus, uncertainty over the probable payouts from new product development becomes more acute.

2. Government regulation. This is discussed in some length in the next section of this chapter.

3. Changes in the tax law. Two schools of thought emerged on the subject of possible effects of changes in the tax law upon the rate of technological innovation. The first of these held that recent changes in the tax law (most notably the increased taxation of capital gains) have had a depressing effect upon the rate of innovation by reducing the potential after-tax returns to successful innovation. It was held that under the present tax structure, the potential gains to shareholders, in particular of small firms engaged in new product innovation, are not commensurate with the risks involved in holding shares in such firms. A second school held that it is not the magnitude of taxation but the effects of a constantly changing tax code that inhibit innovation. The argument was that the possibility of further changes in the future increases the uncertainty associated with new product development.

4. Industry structure. The argument was made that in certain industries, at least, industry structure may lead to patterns of firm conduct that discourage new innovation. This might be the case in industries where interdependence among firms is high. It was felt that in such industries the rate of defensive investment (investment to protect market share) would be high and that investment in the development of new products would be correspondingly decreased.

5. Government antitrust policy. In some industries, it was held, the fear of antitrust proceedings being brought

against large firms inhibits these firms from introducing new products that might expand the market power of the firms.

6. Reduced rates of government funding of R&D. It was noted that real government expenditures on R&D have not increased markedly in recent years and have actually declined as a percent of gross national product (GNP). (See Table 4.) Particularly affected have been programs in R&D in universities.[2] It was noted that governmental cutbacks in expenditures upon R&D, however, have fallen most in the space- and defense-related sectors, and it was questioned by some participants whether such cutbacks ought to have much effect upon the rate of innovation in commercial technology.

3. WHAT SHOULD BE THE ROLE OF THE FEDERAL GOVERNMENT IN PROMOTING TECHNOLOGICAL INNOVATION?

Much of the discussion of why there might currently be a laggard rate of technological innovation in the U.S. economy singled out federal government policies as a possible cause of the laggard performance. In this context, it was

TABLE 4 Federal Funding for Research and Development ($ millions)

Year	Funds in Current Dollars	Funds in 1967 Constant Dollars[a]
1964	12,553	13,200
1968	14,952	14,400
1970	14,764	12,800
1972	15,875	12,500
1974[b]	16,955	12,100
1975[b]	18,160	NA

NA = not applicable.
[a]Estimated by the author using GNP deflator.
[b]Estimated.

SOURCE: *National Patterns of R and D Resources: Funds and Manpower in the United States* (Washington, D.C.: National Science Foundation, 1975).

asked, what should be the role of the government in promoting technological innovation?

It was felt that there are three components to this question. First, under what circumstances is it appropriate for the federal government directly to fund R&D? Second, in what ways can the federal government indirectly encourage private investment in R&D? And, third, how can the federal government ensure that its policies do not unintentionally discourage technological innovation? Each of these is taken up in turn.

Under What Circumstances Should the Federal Government Directly Fund R&D?

Discussion of this question, it might be noted at the outset, was limited to what role the federal government might have in funding nonmilitary R&D. There was little discussion of the need for government funding of military R&D, outside of a general agreement that such funding is necessary but has been at times excessive in relation to the benefits derived.

In the first chapter of this report, the argument is presented that under a private market system, investment in the creation of new technology occurs only when private firms have an economic incentive to make such an investment. The argument was made, however, that the social value of investment in new technology can exceed the private return to the investment, leading to a net investment in new technology that would be less than the socially desirable level of investment. This suggests the possibility that public funding of investment in new technology might be desirable.

Few workshop participants questioned the logic of the possibility, but at the same time the feeling was strong that extensive federal funding of applied R&D at the industrial level was not generally desirable. Federal funding of basic research and engineering and of the educational system, however, was seen as being both necessary and desirable.

The lack of enthusiasm for direct federal funding of commercial R&D centered mostly about institutional problems with such funding. It was felt that there must be a strong coupling between the marketplace and the R&D activity in order that investment in R&D be allocated to the most promising pursuits. If the federal government were to fund R&D in industrial firms, the government would be placed in a position of having to decide how the R&D funds would be spent. In the opinion of several workshop participants placing this decision in government hands would decouple the decision of how to allocate R&D funds from the marketplace, resulting in inefficient usage of such funds. The strong feeling was that the government is not well equipped to make this sort of decision.

It was further felt by some participants that because technology developed at public expense becomes property in the public domain, there would be little incentive for the firm that created such technology to exploit it to its full commercial potential. There was not total agreement on this point, however. It was pointed out that the net social benefits resulting from allowing all firms to have the right to exploit such technology commercially might exceed the social benefits resulting from allowing the original creator of the technology to maximize its own returns from proprietary use of the technology.

There was agreement that there might be special cases in which direct government funding of commercial R&D might be warranted, in spite of the institutional problems. Generally, if the magnitude of investment required to develop a particular technology were to be so great that no private firm would willingly undertake the investment, but the social benefits of the investment were anticipated to be greater than the social costs, it would be desirable for the investment to be publicly funded. However, it was felt that such cases are quite exceptional and that much caution should be exercised before such funding is to be undertaken.

Most participants agreed that federal funding of basic scientific research is necessary and appropriate. Scientific reserach is, in the opinion of most, the ultimate fountainhead of technological innovation. The major problem is: How much basic scientific research is necessary? The question is difficult to answer, for at least three reasons. First, not all scientific research immediately leads to direct commercial application. (But, as any scientist will quickly point out, neither should it necessarily have to; the purpose of scientific research is to add to the body of scientific knowledge.) Second, the lead time between scientific discovery and commercial application is often long. Third, and perhaps most important for this discussion, the ultimate technological benefits from scientific discovery are uncertain and unpredictable. Many participants (but not all) felt that basic scientific research is presently underfunded in the United States and that the present trend toward spending a reduced percentage of national income on scientific research should be reversed.

The major appropriate domain for federal funding of the creation of technology was considered to be in scientific and engineering education. Several participants believed that support for applied engineering in particular is inadequate. It was noted that few engineering departments in U.S. universities now offer extensive curricula in basic production engineering or plant design; instead, the curricula are largely oriented toward the engineering sciences. This orientation was believed to have been the result in large measure of extensive federal support at the university level in past years for the development of high technologies oriented toward space and defense programs but much less support for applied engineering oriented toward commercial technology. The net result, in the eyes of several workshop participants, has been that the U.S. engineering educational system has produced more graduates oriented toward the high or exotic technologies than the economy can usefully employ but a shortage of engineering graduates equipped to innovate in manufacturing technology, the technology that would result in productivity advances.

The participants generally agreed that federal funding for engineering education should be increased and that increased emphasis should be placed upon applied engineering skills. It was also proposed that the federal government might explore ways in which existing engineering talent in this country might be retrained, so that an engineer who finds opportunities diminishing in his particular area of specialty might more easily be able to change specialties. It was noted that examples abound of Ph.D.-level engineers who are unable to find employment that makes use of their specialty. It was felt that efforts must be increased to enable such persons to acquire new skills and redirect their talents toward specialties that are in greater demand.

It was felt by some participants that differences in emphasis in the educational process could account for at least some of the differences in productivity growth among the OECD nations. Japan, Germany, Switzerland, and Sweden, it was felt, are nations that have put a heavy emphasis upon training engineers to serve in the commercial sectors of their economies, and these nations have been pioneers in the development of new manufacturing technologies. These nations have experienced the most rapid growth in productivity of the world's nations, and it was felt that this could in part be accounted for by improvements in the quality of capital resulting from the types of engineering skills present in these economies. Although statistical data (other than inferential data) are not available regarding the quality of capital among different nations, the opinion of some workshop participants was that the average capital stock of nations such as Germany, Japan, Sweden, and Switzerland is superior in quality to that of the United States. Furthermore, the opinion was that new capital stock being added in these countries was superior in quality to that of the United States.

An outgrowth of this discussion was a suggestion by several participants that there might be some utility in creating data on the relative age and quality of the stock of capital goods of the major industrial nations by industrial sector. It was believed that data of this sort could be useful in ascertaining the major determinants of differences among nations in rates of improvement of productivity. In particular, it could be determined whether age, obsolescence, or quality of the U.S. capital stock is a significant factor behind the reported poor performance of the United States in productivity advance. It was also suggested that such data could be used to determine whether or not differences in the relative quality of U.S. capital stock from that of other nations were more or less apparent in the "multinational" industries. This, it was believed, could be useful in determining whether or not technology transfer abroad by U.S. multinational corporations could in any way be held accountable for slow advances in U.S. productivity.

One participant noted, however, that the primary difference between technological innovation in the United States and that abroad was that innovative activity in the United States historically has emphasized the development of new products (see evidence in Chapter 2) while innovative activity in Germany and Japan has emphasized making existing products more cheaply. It was felt that productivity measures understated the relative economic performance of the United States in relation to that of major OECD nations, because productivity per se does not measure the utility gained from new product introduction.

How Might the Federal Government Indirectly Encourage Technological Innovation?

The strongest consensus on this question was that not enough is presently known about the ways in which the federal government, advertently and inadvertently, affects the rate of technological innovation. It was felt by the workshop that research into this area is a matter requiring urgent attention and that study of this topic should be given high priority by the National Science Foundation.

Several positive recommendations emerged from discussion of this question:

First, it was felt by some participants that U.S. tax policy has a substantial impact on the rate of technological innovation. Several workshop participants believed, for example, that some combination of tax measures could increase the incentive for firms to invest in the creation of new technology. For example, rapid write-offs of expenditures on R&D would increase the after-tax present value of such expenditures and thus presumably provide a greater incentive for firms to make such expenditures. A reduced capital gains tax might increase the incentives for individuals to invest in new ventures to develop and commercialize new technologies.

Not all participants believed that favorable taxation treatment was the correct route to stimulate investment in new technology, however. The point was made that tax breaks for investment in technology are in effect a public subsidy for this investment, but that the returns from the subsidy go largely to private individuals.

There was agreement that too little is known about the effects of tax policy upon the rate of technological innovation. It was felt that further study of these effects is warranted.

It was proposed that the federal government could indirectly encourage technological innovation by amending the antitrust laws to permit firms within an industry to create joint ventures to conduct at least some kinds of basic research, the results of which could be shared among firms. It was felt by some participants that such joint ventures would reduce the riskiness, at the level of the firm, of certain types of investment in new technology. It was felt by the workshop that this was a matter warranting further study.

There was agreement by most of the workshop participants that the most effective indirect means by which the federal government could act to increase the rate of technological innovation would be to stimulate the rate of new capital formation. It was noted that empirical studies have shown that innovation rates tend to be highest in industries where there is a high rate of capital investment. (See Chapter 2 of this report.) It was further noted that capital investment rates in the United States have been low throughout the 1970's.

How Can the Federal Government Ensure that its Policies
Do Not Adversely Affect the Rate of Technological
Innovation?

Numerous participants in the workshop believed that many federal government policies during the last decade have, perhaps inadvertently, reduced private incentives to invest in the creation of new technology. Such policies, it was felt, included monetary policies (it was felt that fluctuations in the rate of interest increased uncertainty over paybacks to investment, including investment in research), tax policies, and regulatory policies.

The latter were singled out as a special example of how federal policies could adversely affect the rate of technological innovation. Although it was recognized that this is a topic in which much has to be learned, several participants at the workshop, especially those with extensive industrial R&D experience, expressed the view that federal regulation in the United States has had a selectively depressing effect on the rate of innovation of U.S. industry and has reduced the ability of U.S. firms to compete effectively internationally. It was felt that this was true for regulation designed to protect the physical environment, regulation designed to enhance human health and safety, product standards regulation (such as that of the Food and Drug Administration (FDA) in the drug industry), and price regulation.

Evaluation of the economic consequences of regulation designed to protect the environment poses special problems that illustrate the difficulties of discussing regulation. Pollution of air, water, and land resources results in real but often intangible costs to society, but the costs generally do not appear in any accountant's ledger. In the economist's jargon, these costs are "externalities." Efforts by firms to abate pollution caused by their operations do pose tangible costs to the firms, costs that must ultimately be borne by the consumers of the firms' products. Two problems are associated with attempting to evaluate the economic impact of this regulation. The first lies in attempting to determine whether the social gains of pollution abatement are commensurate with the social costs. This determination is made very difficult because there exist no analytical tools capable of assessing exactly what are the magnitudes of the intangible components of the gains and losses. The second problem is that the costs of regulation are not borne evenly across industries. To the extent that uneven sharing of costs causes the relative prices of goods to shift, the result is an allocation of existing resources different from that which would occur in the absence of regulation. The magnitude of this reallocation and the extent to which it reduces consumer utility are extremely difficult to measure. The critical question to be answered is as follows: Is the total social cost of regulation including reallocation of existing resources greater than or less than the utility gained from reduced pollution?

As was stated, some workshop participants believed that environmental regulation has a negative impact on technological innovation. The argument was that regulation necessitates that R&D funds, which otherwise might be allocated to developing new products or to increasing the efficiency of the manufacture of existing products, instead be diverted to the development of products and processes that meet the standards imposed by the regulation. From a societal point of view, this may or may not be a bad thing, depending upon which the society values more: new or cheaper products or a less befouled physical environment. From the point of view of the firm, however, there is little or no opportunity for the capture of a tangible benefit from the investment in R&D to meet the regulation, but there are positive financial returns from other types of R&D activities. Therefore the firm would, of course, choose the latter activity.

The view that environmental regulation necessarily causes a diversion of innovative efforts from tangibly productive to nonproductive uses, however, did not go unchallenged. The point was made that the reverse might occur. In the automotive industry, for example, federally mandated emission control and safety requirements have forced the major producers to conduct R&D to make their products cleaner and safer, R&D viewed by some participants as a relatively productive use of resources. At the same time, there is little evidence to suggest that these efforts have displaced the more traditional R&D activities of the automotive firms. Thus, it was questioned whether or not resources allocated to the meeting of mandated regulations necessarily displace those that would be used to generate newer products or to reduce costs of existing products. It was further reported that the tentative conclusions of some recent studies of the overall impact of environmental regulation upon the costs of production of goods indicate that compliance with regulations may not always result in higher costs. In some cases, the need to meet clear air and water standards has caused plants to be redesigned so that factor productivity is enhanced as well as effluents reduced. It is not known whether such cases are highly exceptional or not.

Concerns about environmental regulation, as voiced by industrialists, often reflect the fact that industries are not perfectly competitive and that environmental regulations are not uniform across industries. Thus, for example, small paper companies operating in the northeastern United States, often already operating at a cost disadvantage with respect to larger competitors located in the southeastern and northwestern areas of the country, are faced with stiffer demands to abate pollution than are some of their larger competitors. For many of these small companies, the cost of meeting environmental regulation is reported to be too great to allow them to remain in business. They simply cannot pass the costs on to customers without losing customers to other firms. In this example, and in other similar cases, it is a matter of argument whether it might be better to allow marginal producers to go out of business or to allow them to continue to emit effluents at existing levels.

Outside of the domain of environmental regulation, the effects upon innovation of other types of regulation were discussed. One participant noted that regulatory activities of the FDA are widely perceived to have slowed down the rate of introduction of new pharmaceutical products. (See the discussion in Chapter 2.) Two reasons for this are commonly noted. First, the development costs associated with creating new pharmaceutical products have been rising at an annual compound rate of about 30 percent a year since 1967; at least some of this rise is accounted for by increases in federally mandated clinical trials and more elaborate toxicology studies required before a new drug is approved to be placed on the market. It was noted that the per annum rise in development costs has been about 30 percent over the past decade, a figure that far exceeds the compound growth rate of the revenues of the pharmaceutical firms. Second, the amount of time that passes between the submission of a new drug application to the FDA and the approval of the application increased dramatically during the 1960's (but has decreased since then). It was noted that the rate of increase in development costs of a new drug application has been rising somewhat faster in the United States than abroad. (See Table 5.)

A common complaint voiced about regulation in the United States by industrial leaders is that it is not the regulation itself that necessarily is onerous but the manner

TABLE 5 Average Cost of New Drug Development and Average Length of Time Required for Approval of New Drug Application, United States and Overseas

	1962	1969	1972
Average cost of new drug development (millions of dollars)			
United States	1.2	3.0	11.5
Overseas[a]	0.9	2.1	7.5
Average time for approval of new drug application (months)			
United States	6	40	--[b]
Overseas[a]	6	9	16

[a]United Kingdom, Holland, Sweden, France, and West Germany.
[b]After 1969, changes were made in the FDA that resulted in a major shortening of the approval time in some cases but not in others. The length of approval time depends upon the particular drug and upon the division of the FDA having responsibility for the approval.

SOURCE: *Research Management*, March 1974, pp. 18-19.

and speed in which it is implemented. The opinion was expressed at the workshop that inconsistency and unpredictability in the application of regulatory standards is a common feature of the regulatory process. Thus, it was felt that the uncertainties posed by regulation can have an inhibiting effect upon both rates of new capital investment and rates of investment in innovative efforts, whereas there would be less inhibiting effect were the regulation to be consistently applied and changes implemented over a longer period of time.

Virtually all participants of the workshop concurred that regulatory procedures over a wide range of activities need revamping and, in some cases, reduction. Many felt that many of the goals of regulation could be accomplished without having a depressing effect on the performance of the regulated industries if the regulatory procedures were to be better administered. Better administration, it was felt, would result if regulatory agencies would set objectives and then devise regulations that allow for the greatest possible flexibility in meeting these objectives.

Too often, it was felt, regulatory agencies decide upon an objective and issue a set of highly specific instructions on exactly how the objective is to be met, only later to change the objective. This, it was felt, penalizes firms that seek to comply with the original objective. Compliance often results in hefty expenditures by the complying firm, and an abrupt change of objective can result in the expenditures, for all practical purposes, being wasted.

A commonly heard statement about regulatory administration procedures is that the amount of documentation associated with regulatory compliance is excessive. This complaint is most often made in conjunction with the documentation that must accompany a new drug application in the pharmaceutical industry, but it is also made in the context of Environmental Protection Agency (EPA) and Occupational Safety and Health Administration regulation. The opinion was expressed at the workshop that the sheer volume of required documentation far surpasses the abilities of decision makers in either the regulatory agencies or the regulated firms to digest effectively the data contained therein. It was recommended that simplification of documentation and reporting requirements could contribute significantly to the effectiveness of regulatory administration.

On the whole, it was felt by the workshop that not enough is known about the effects of regulation on the economy. It was recommended that the federal government give a high priority to the study of this issue, preferably by independent researchers, and in particular to sponsor independent research on the impact of regulations upon technological innovation.

It was also recommended that regulatory efforts by the federal government should be pursued only after careful cost-benefit analyses of the economic consequences of the regulations had been achieved. It was stressed that such cost-benefit analyses should be reviewed by representatives of all major interested parties, including industrialists, labor, consumers, and environmental protection groups. It was felt that such analyses, properly done and properly reviewed, could be used to determine the true costs and benefits of regulation. This could in turn be useful in determining what types of regulation are necessary and effective and what types are unnecessary or ineffective.

The point was made by one workshop participant that regulatory agencies operate under charters that often provide no incentive to improve the efficiency or efficacy of their operation. This results from the incompleteness of their charters; for example, the FDA is given responsibility for ensuring the safety of drugs, but does not take equal responsibility for providing the public with a flow of new drugs. In such instances, it was felt that the internal environment within these agencies lends itself to an inexpeditious administering of the regulatory process. It was suggested that regulatory agencies should be subjected to a continuous review to assure that they are meeting reasonable standards of efficiency.

The point was raised at the workshop that the costs of meeting regulatory requirements add costs to goods that are manufactured in the United States, costs that are not borne by foreign competitors. This, it was argued, distorts the terms of trade of the United States in international markets, making exports from the United States relatively more expensive and imports into the United States relatively less expensive than would be the case in the absence of regulation.

The most ready answer to this problem would be for the United States to place some sort of an equalizing tax on imports and to grant a rebate to exports, each calculated to offset the extra costs borne by U.S. producers but not by foreign producers to meet regulatory requirements. The problems posed by this "ready answer," however, are manifold. The magnitude of the tax or rebate would be very difficult to calculate and presumably would vary depending upon the final destination of the export or the origin of the import. The application of differing rates of tax or rebate to different nations could be construed as being in violation of the "most favored nation" provision of the General Agreement on Tariffs and Trade (GATT). Indeed, such actions could be viewed as being in violation of the general spirit of GATT. Imposition of taxes could be seen as a raise in U.S. tariffs, and granting of rebates as an export subsidy.

Whether or not an import tax and export rebate program is practical needs to be explored in depth. In particular, an evaluation should be made of what, if any, damage would ensue to the world trading system were an effort made to implement such a program.

4. IS THERE A BASIS FOR RESTRICTION OF U.S. TECHNOLOGY TRANSFER ABROAD?

Whether or not the U.S. government should (or even can) act to restrict technology transfer abroad is, in the final analysis, a judgmental matter. Two questions are pertinent. First, does the United States have unique possession of technologies that would be in the best national interest not to transfer to other nations?[3] Second, if such technologies do exist, is there a feasible means to restrict international transfer of these?

As was suggested at the beginning of this chapter, there was no genuine consensus on the first of these questions. Most workshop participants did believe that the United States does in the 1970's still have some sort of technological advantage over other nations in some types of technology. A few participants believed that the United States should attempt to hold on to whatever technological lead it has by whatever means possible. Most participants, however, believed that the only meaningful way in which the United States could continue to be the most technologically advanced of nations was to continue to invest in the creation of new technology.

Efforts to restrict the export of U.S. commercial technology, it was believed by most participants, would be undesirable for at least two reasons.

First, it was felt that no such effort would work. It was noted that efforts historically by other nations to contain technology at home had generally failed miserably. Technology is knowledge, and the spread of knowledge is hard to prevent.

Second, and more importantly, it was felt that the United States is a nation that has throughout its history been a strong advocate of free enterprise. In the postwar era, the United States has been a leader in the movement to liberalize restrictions on international trade. It was felt that it would be contrary to the principles of free enterprise and free trade were the United States to advocate restrictions on the international transfer of commercial technology. Such an advocacy would be akin to economic nationalism and would bring with it all of the worst aspects of neomercantilism. The spectacle of the United States embarking upon such an advocacy was not something that the majority of the workshop wished to observe.

NOTES

1. For one industrialist's view of this issue, see the statement of Ralph Landau to the Joint Economic Committee of the Congress of the United States, Tuesday, October 2, 1974 (U.S. Government Printing Office Document 49-914).
2. Changes in levels and priorities of federal funding of R&D have affected universities in a number of ways. See B. L. R. Smith and J. J. Karlesky, *The State of Academic Science, The Universities in the Nation's Research Effort* (New Rochelle, N.Y.: Change Magazine Press, 1977).
3. Excluded from this discussion were technologies that are highly sensitive from a military standpoint. All workshop participants believed that there should be strict controls over the international transfer of these. Many participants felt also that the international transfer of certain categories of technology should be restricted even if the United States does not hold a monopoly position over them; for example, most participants felt there should be a strong multilateral control over certain categories of nuclear technology. The technologies under discussion were largely commercial technologies not having major military dimensions.

4 THE INTERNATIONAL TRANSFER OF TECHNOLOGY, INTERNATIONAL TRADE, AND INTERNATIONAL INVESTMENT: THE POINT OF VIEW OF U.S. ORGANIZED LABOR

1. THE LABOR POSITION

It is a fact that the organized U.S. labor movement in 1977 is in support of legislation to restrict foreign direct investment by U.S. firms on the grounds that many such investments "export jobs" overseas. At the same time, certain individual U.S. unions are calling for measures to restrict imports of products that compete in domestic markets with products produced by the U.S. workers whom these unions represent. In short, the overall position of U.S. organized labor in 1977 is somewhat protectionist.

United States organized labor, according to its spokesmen, has not always held a protectionist position. As early as the late 1920's, labor was on record as opposing the Smoot-Hawley Act. Throughout most of the post-World War II era, the U.S. labor movement has lent its support to efforts by the U.S. government to liberalize international trade and investment rather than to restrict it. United States governmental aid programs from the time of the Marshall Plan have been supported by labor, and labor officially supports the latest congressional appropriation for the U.S. Agency for International Development. The Kennedy Round of Tariff Reductions in the early 1960's received the official support of organized labor. Although there have always been some reservations expressed by labor leaders over trade liberalization (particularly by leaders of unions representing workers employed in labor-intensive industries) and certain individual unions historically have consistently sought trade restrictions, the labor movement prior to the late 1960's generally endorsed efforts to liberalize trade.

Labor leaders today, when reviewing the reasons for labor's shift in position, cite the major reason for labor's support of free trade during the 1950's and 1960's as the "long lead" of the United States in technology.[1] The long lead, it is argued, gave the United States a "comparative advantage" in trade with respect to its major trading partners. The implications of this comparative advantage, as articulated by labor leaders, were that as impediments to trade were removed, U.S. exports would rise, creating more jobs for U.S. workers.

Labor representatives present at the Woods Hole Workshop cited a declining U.S. lead in technology to be a prime reason for labor's increasingly protectionist position. It was their position that U.S. corporations, by investing abroad to take advantage of lower labor costs overseas, were "trading away" the comparative advantage in technology held by the United States. They further noted that U.S. corporations have been making upward of 25 percent of their total capital investment overseas in recent years, and it was asserted that as a result the U.S. economy loses employment opportunities. (Table 6 indicates the total domestic U.S. capital investment by the private sector in recent years and the total overseas capital investment by U.S. private companies during the same years, broken down by major industry.) The long-term health of the economy, in labor's view, is further endangered by the transfer of managerial know-how to overseas subsidiaries of U.S. corporations and the shipping of advanced machinery abroad.

The point was argued by labor leaders that transfer of U.S. technology began in the low-skill, labor-intensive industries, such as shoes, textiles, simple electronics assembly, and leather.[2] The feeling was that the transfer was progressing to higher-skill-level industries, such as the assembly of complex electronic components and turbines.

It was recognized by labor representatives that newly developed technology is not something that can be forever contained in the United States. However, it was argued that the U.S. government pursues policies that actively promote the outflow of U.S. technology and, furthermore, that governments of other nations pursue policies to protect their domestic industries that the U.S. government fails to match. In the former category, the two U.S. governmental policies most scored by labor representatives were tax and tariff policies.

The U.S. tax code, according to labor, contains numerous provisions that make it more profitable for U.S. corporations to invest abroad than to invest at home. The major provisions objected to by labor are relatively well known: tax credit for foreign income taxes paid or deemed paid

TABLE 6 Capital Investment by Private Sector in the United States

Industry	(A) Plant and Equipment Investment Abroad by Affiliates of U.S. Corporations ($ millions)					(B) Plant and Equipment Expenditures in the United States by the Private Sector ($ millions)						(C)
	1971	1972	1973	1974	1975	Average, 1971-1975	1971	1972	1973	1974	1975	Average, 1971-1975	Average,[a] 1971-1975
Food and beverage	324	364	547	530	710	495	2,690	2,550	3,110	3,250	3,920	3,100	0.16
Paper products	524	586	621	807	719	651	1,250	1,380	1,860	2,580	3,330	2,080	0.31
Petroleum	4,959	5,350	6,637	8,765	9,804	7,103	5,850	5,250	5,450	8,000	10,510	7,010	1.01
Chemicals and pharmaceuticals	1,397	1,349	1,322	1,706	2,597	1,674	3,440	3,450	4,460	5,690	6,670	4,740	0.35
Rubber products	232	413	335	273	428	336	840	1,080	1,560	1,470	1,000	1,190	0.28
Metals	840	702	722	687	712	732	2,780	2,750	3,430	2,120	5,730	3,370	0.22
Nonelectrical machinery	1,794	1,716	2,619	2,869	2,753	2,350	2,800	2,900	3,420	4,420	5,090	3,730	0.63
Electrical machinery	513	586	883	895	905	756	2,140	2,390	2,840	2,970	2,530	2,570	0.29
Transportation equipment	889	855	1,109	914	1,279	1,009	2,130	2,530	3,120	3,750	3,430	2,990	0.34
Other	593	552	758	871	1,004	756	6,060	7,070	8,780	8,930	8,400	7,850	0.10
Total, petroleum, and manufacturing industries	12,065	12,473	15,553	18,317	20,911	15,864	29,980	31,350	38,120	43,180	50,610	38,560	0.41
Total, excluding petroleum	7,106	7,123	8,916	9,552	11,107	8,761	24,130	26,100	32,670	35,180	40,100	31,640	0.28

[a] (A) ÷ (B).

SOURCE: U.S. Department of Commerce, *Survey of Current Business* (various issues).

(full tax credit is possible if the weighted average of foreign tax rates is less than the U.S. rate and the reporting corporation elects the "overall limitation" method of reporting), deferral of U.S. income taxes on unrepatriated foreign income, and special provisions for reduced taxation of income originating from developing nations.[3] The general feeling of the labor representatives at the workshop was that (1) tax credits should be eliminated and foreign tax should be treated as a deductible expense rather than as a credit, (2) deferral should be eliminated, and (3) income from developing nations should not be given preferential tax treatment by the U.S. government. Reactions of the workshop to these suggestions are reported in the next section of this chapter.

Certain U.S. tariff policies were scored by labor representatives. Their principal complaint was that foreign countries often charge relatively high tariffs on imports of products on which the United States charges lower tariffs. The complaint was largely directed toward perceived tariff imbalances between the United States and the larger developing nations, although it was noted that some imbalances exist between the United States and OECD trading partners.[4] Japan, in particular, was noted as one OECD nation that traditionally has refused to allow imports to come into her home market on as favorable terms as those on which Japanese firms are able to export to foreign markets. It was noted, however, that during the 1970's, Japan has moved rapidly to reduce barriers to importation into the Japanese market.

An example given involved the light aircraft trade between the United States and Brazil. Until the early 1970's, Brazil was a leading purchaser of light aircraft manufactured in the United States.[5] However, in 1971, the Brazilian government decided to create a domestic light aircraft industry, and prohibitive tariffs were enacted on the importation of light aircraft into Brazil. Not having available domestically the know-how required to produce light aircraft, the Brazilian government invited the Piper Aircraft Corporation to set up shop in Brazil, and in 1974 Piper Aviacao do Brasil, Ltd., was created to manufacture light airplanes for the Brazilian market as a joint venture with Empresa Brasileira de Aeronautica, a mixed state-private Brazilian enterprise. In 1976, Piper Aviacao do Brasil supplied about 75 percent of the domestic Brazilian market, while the U.S. export market share fell from almost 100 percent in 1970 to less than 1 percent in 1976. Furthermore, in 1976, the Brazilian

Piper subsidiary had begun to export aircraft to Uruguay, Chile, Peru, Colombia, Venezuela, and some African nations. Attempts were being made by the Brazilians to sell aircraft in the United States. By labor's estimate, the creation of a light aircraft industry in Brazil has cost the U.S. aircraft industry several hundred jobs. Of special concern to U.S. labor was that in 1976 Brazil effectively embargoed the importation of any aircraft into Brazil. In that year, no new import permits were issued by the Brazilian government at all. (Imports of light aircraft into Brazil were in fact recorded, but these apparently were made on the basis of previously issued permits.)

It was expressed by the labor representatives that foreign governments, especially those in Western Europe, pursue a wide variety of policies to protect the domestic interests of their labor and pursue these policies much more vigorously than does the U.S. government. Thus, for example, if a company in Sweden or Western Germany wishes to close a plant and lay off the workers, it must generally obtain governmental permission to do so. In the United States no such permission is needed: if closure of a plant is believed by a U.S. company to be in the best economic interests of that company, it can close the plant irrespective of the external social cost of doing so. The strong feeling expressed by labor representatives at the Woods Hole Workshop was that labor should have some input to a company's decision to close a plant.

The labor view is that many overseas governments give to local industries a wide variety of domestic and export subsidies. Some of these subsidies are overt, but many are indirect and difficult to uncover. The net effect of unilateral foreign subsidies, it was felt, is for the benefits of trade to be skewed against the United States. United States labor believes that the policy of the U.S. government generally should be to persuade foreign governments to stop subsidizing their exporting industries and to prohibit further increases in U.S. imports of subsidized goods until the subsidies are ended.

2. REACTION OF THE WORKSHOP TO LABOR'S POSITION ON THE TRANSFER OF TECHNOLOGY

Labor's concerns with the transfer of technology, as expressed at the workshop, could be roughly broken down into two categories. The first was that U.S. transfer of

technology, primarily via foreign direct investment but
also via the export of machinery embodying advanced technology, has some long-term damaging effect on the U.S.
economy as a whole. Discussion of this issue is reported
in Appendix A. The second was that U.S. labor is saddled
with the costs of U.S. trade and investment policies, while
the gains are reaped by other sectors of U.S. society.
This section of the report addresses primarily the workshop's reaction to the second category of labor's concerns.

The general reaction of the nonlabor representatives
at the workshop to the labor position was one of somewhat
qualified sympathy. It was felt by most participants that
there is some validity in labor's position, which requires
some sort of redressing. The specific policy changes
sought by labor, however, were not generally supported by
the group as a whole.

It was noted that the strongest reason for labor's feelings against liberal trade and investment policies is the
fear of job loss.[6] The disruptive effect of job loss upon
a worker, his family, and, in some cases, his community is
difficult to measure. The older and more established the
worker, the greater the trauma associated with job loss
is likely to be. Even if alternative employment possibilities are open to a displaced worker, such possibilities
often involve retraining or relocation. Unlike the executive- or professional-class worker, who often changes jobs
or locations in response to greater perceived opportunities,
the displaced blue-collar worker does not necessarily expect to benefit from a change of job. More often, the
expectation is one of hardship or loss.

The social cost of adjustment to job loss is often not
fully accounted for by economists when analyzing the impact
of international trade and investment upon the economy. In
making such analyses, economists often assume (for analytic
convenience) that there is perfect, long-term factor mobility within the economy and that factor prices move in
harmony with the marginal productivities of the factors.
Thus, in the economist's view, the net effects of international trade and investment are changes in national
income and the distribution of this income. In a large,
capital-rich economy such as the United States, the economist's analysis indicates that unimpeded international
trade and investment would increase total national income
over what income would be in the absence of this trade
and investment. However, factor price equalization implies
a rise in the relative factor price of the relatively
abundant factor (capital) to that of the relatively scarce

factor (labor). Thus, a redistribution of income from labor to holders of capital is believed to occur. The economist's analysis suggests that, in the long-term at least, the major impact of free international trade and investment upon labor is a wage rate lower than that which would have been realized had the same amount of capital been invested at home.[7]

One participant, in response to this analysis, noted that the high-technology sector of the U.S. economy is its "cutting edge," and although jobs are lost in some of the less advanced industries, other jobs are being created in such higher-technology industries, where investment must flow abroad to maintain market share. Such foreign investment often requires more intermediate or supply exports from the United States than were previously required.

The long-term redistribution of income from labor to capital as a consequence of international trade and investment is not the least of organized labor's worries, but, in the view of the workshop, neither is it at the front of the list. The assumptions of factor mobility and flexible factor prices, made for analytic convenience by the economist, mask the more immediate problem of job security. The economist notes that there are long-term gains from international trade and investment and worries about how these gains can be equitably distributed. Of more immediate concern to the worker (and, hence, to the labor movement) is the short-term problem of unemployment and adjustment to this.

The type of worker who becomes unemployed because of imports warrants some mention. A 1976 study compares a sample of workers who were laid off their jobs because of imports with a sample of workers who were laid off for other reasons.[8] The study showed that import-affected workers tend to be older and less well educated than the non-import-affected workers. Generally, the jobs lost on account of imports are likely to be categorized as semi-skilled or unskilled. The import-affected workers were found to have had accumulated a much greater number of years of job tenure before being laid off than had the non-import-affected workers, and the import-affected workers had a much more difficult time in finding new employment than did the non-import-affected workers. A conclusion of the study was that the type of worker who was laid off on account of imports was one who would be expected to have a high degree of difficulty in adjusting to the layoff.

The major issue raised by this study is how the long-run

gains from international trade and investment can be captured by the nation without posing undue hardship upon any sector of the society, especially job loss to immobile sectors of the work force.

It was pointed out by one workshop participant that with the passage of the Trade Act of 1974, the United States initiated a program of "adjustment assistance" designed to minimize the disruptive effects of liberalized trade. The 1974 program superseded the Trade Adjustment Assistance Program of the Trade Expansion Act of 1962. The idea of the program is that workers who lose their jobs because of imports can receive up to 70 percent of their weekly wage in combined unemployment insurance and Trade Readjustment Allowance (TRA) payments.[9] To be eligible for adjustment assistance, affected workers must submit a petition to the International Labor Division of the U.S. Department of Labor within 1 year of being laid off. The petition must be signed by at least three laid-off workers or by a union official or authorized representative. The International Labor Division can approve the petition if three criteria are met: (1) there has been a substantial increase in unemployment in the industry in which the laid-off workers were employed; (2) there has been a substantial decrease in industry sales and production; and (3) there has been a substantial increase in imports in the industry.[10] The criteria are designed to ensure that only those persons laid off work because of import competition can receive TRA. Whether or not a particular group of laid-off workers do, in fact, qualify for TRA under the three criteria is, in the final analysis, a judgmental matter that must be decided by the International Labor Division.

Under the readjustment assistance program, a qualifying worker can receive governmentally subsidized training if such training can help him obtain employment that otherwise would be unavailable. To receive training, the worker must indicate to the Employment Services (ES) Division of the Department of Labor his intention to receive training, and such indication must be received by ES within 18 weeks of the date of the filing of the petition for TRA.

At the time of the writing of this report, there were no published government data on the extent of the TRA program under the 1974 act. A Labor Department official in the New England office indicated that, in very rough figures, about 8,000 workers in New England qualified for TRA and that these received about $10 million in benefits during 1975 and 1976. Of these 8,000 TRA recipients, over half returned to work before maximum TRA benefits were paid. About 5 percent of the total received training, and an additional 5 percent were placed in new jobs by the ES, which involved on-the-job training.

Under the 1962 Trade Act Adjustment Assistance Program, a total of 100,546 workers applied for adjustment assistance, of which 44,849 received assistance and 55,697 were denied assistance. (See Table 7.) It is estimated that the total payments under this program amounted to $71.5 million, or about $1,600 per worker.[11]

Labor representatives at the workshop commented on the TRA program. Most of the commentary was unfavorable. It was argued that TRA criteria for qualification are rigidly

TABLE 7 Distribution of Trade Adjustment Assistance Petitions, by Industry, October 1962 through May 1974

Industry Category	SIC[a]	Petitions		Workers	
		Accepted	Denied	Accepted	Denied
Metal mining	10	0	1	0	650
Food and kindred	20	0	1	0	163
Textiles	22	6	18	2,900	10,876
Apparel	23	0	3	0	1,126
Chemicals	28	0	2	0	1,300
Rubber	30	6	6	4,970	3,073
Leather	31	37	67	11,173	13,617
Stone, clay, glass	32	8	8	2,320	1,920
Primary metals	33	1	7	400	1,982
Fabricated metals	34	3	1	450	200
Nonelectrical machinery	35	2	6	676	3,925
Electrical machinery	36	16	22	15,025	10,205
Transportation equipment	37	2	3	2,150	700
Miscellaneous manufacturing	39	14	4	4,785	5,960

[a]Standard industrial classification.

SOURCE: George R. Neuman, "The Direct Labor Market Effects of the Trade Adjustment Assistance Program: The Evidence from the TAA Survey" (unpublished paper for the U.S. Department of Labor, Washington, D.C., 1977), p. 6.

written and rigidly applied, and, most of all, TRA does not ensure that a worker laid off because of foreign competition can find future employment. It was pointed out that TRA applies only to workers whose jobs are lost on account of imports and not to workers whose jobs are lost because their employer moves its operations overseas.[12]

The rigidity of application of the criteria was illustrated by the case of workers in a shoe last factory which closed down because its customers (shoe factories) had had to close down because of foreign shoe imports. The workers were denied TRA on the grounds that there was no substantial increase in the imports of shoe lasts. This was, of course, the case. The demand for shoe lasts is derived from the demand for shoes, and it was shoes, not shoe lasts, that were being imported.

The expressed feelings of the labor representatives were that TRA is ineffective and does not address the basic problems of labor. The recommendation was that this particular approach be abandoned in favor of direct controls on imports and of overseas investment by U.S. corporations.

Nonlabor participants generally agreed with the position of the labor representatives that the TRA program, as now constituted, is inadequate. There was a general consensus, however, that the correct approach is to expand and modify the program to make it more effective, rather than to abandon it and to impose direct controls on imports or international investment. In particular, it was felt that eligibility requirements should be broadened to include, in addition to workers presently eligible, those workers whose jobs were indirectly jeopardized by imports (such as the shoe last workers) and those workers whose jobs were lost when a firm transferred operations overseas. At the same time, it was recommended that job retraining programs be expanded and that time limits for benefits be extended.[13]

Labor's desire for changes in the taxation of international income engendered discussion by the workshop participants. It was agreed that labor's contention that the U.S. tax code treats domestically earned income differently from foreign-earned income is correct and that some provisions of the tax code probably create a preference for foreign-earned income. However, it was also pointed out that other provisions of the tax code, most notably the investment tax credit, result in a lower tax being paid on domestically sourced income than on foreign-sourced income.

The consensus of the workshop was that the United

States should strive for a position of tax "neutrality," meaning that taxation considerations should neither create an incentive nor a disincentive for the U.S. investor to invest abroad. If the tax system were to be completely neutral, the rank ordering of investment possibilities for any firm would be the same on a pretax basis as on an after-tax basis.

It was recognized that the problem with such a proposal is that there is very little agreement among experts as to what the effects of taxation are upon the decision to invest. Thus, it is extremely difficult to measure what differences might arise in investment patterns as a consequence of changes in the tax code. Studies conducted for the U.S. Treasury on what would be the changes in revenues accruing to the treasury and on the marginal propensity of U.S. investors to invest in the United States and abroad as a result of the elimination of deferrals and credits on foreign income, for example, yield varying estimates, depending upon what the starting assumptions are.[14] To determine what would constitute tax neutrality, the analyst would have to know both exactly how taxation affects the decision to invest and what the investment position of the United States would be in the absence of any tax on investment income (either U.S. or foreign tax). Neither of these is available to the analyst. Compounding this, some analysts believe that the impact of taxation upon the propensity to invest abroad is likely to be different for different industries and perhaps even for different companies. The conclusion of the workshop was that the subject of what constitutes tax neutrality warrants further study.

It was asked if a study of the effects of the tax system as a whole upon the propensity of U.S. firms to invest overseas had even been undertaken. To the best of any participant's knowledge, no such total study has even been done, although numerous studies both within and without the U.S. Treasury Department have been conducted on the effects of specific provisions of the tax code. Furthermore, due to the complexity of the issues and the corresponding need to make simplifying assumptions to reduce the complexities to manageable proportions, it was doubted that a meaningful study could be done. The assumptions, it was felt, usually wind up causing more questions to be asked than are answered.

Most workshop participants expressed a belief that the U.S. tax code should be "neutral" with respect to taxation of overseas versus taxation of domestically sourced income,

and while it was understood that it was difficult even to
define such neutrality and that its attainment would be
still more difficult, few agreed with labor's contention
that elimination of the foreign tax credit would be a move
toward neutrality. Rather, as argued most forceably by
representatives of the business community, such a move would
certainly put a heavy penalty on foreign-sourced income,
causing such income to be subject to a double taxation by
foreign and by U.S. tax authorities. Even if foreign taxes
were treated as allowable business expenses, it was argued,
U.S. corporations would be subject to a very high rate of
taxation on foreign income, a rate higher than that of
their foreign competitors. This would put the U.S. firm
at a competitive disadvantage in overseas activities with
respect to foreign competition.

 Labor's concern over differences in tariff structures
between the United States and certain U.S. trading partners
was felt by many of the workshop to be a valid one. The
problem is that many developing nations see high protective
tariffs in their home markets as an effective means to protect and nurture their fledgling industries. At the same
time, developing nations regard exportation as the only
means by which these industries can become internationally
competitive. Remove the protective tariffs, so the argument goes, and newly established industries in developing
nations run the risk of being put out of business. Remove
the ability of these industries to export, and they will
remain forever inefficient and perpetually dependent upon
the high protective tariffs. In effect, the argument of
the developing nations is that if they are ever to industrialize, their local industries must capture the entirety
of their home markets and a share of an export market as
well.

 Labor's complaint with this argument is that it is the
worker in the industrialized world who must "pay the bill."
Many of the participants at the workshop were sympathetic
to U.S. labor's concern but simultaneously not unsympathetic
to the arguments of the developing nations. Two points
were brought out. First, high protective tariffs are
justified to protect an "infant industry" only for as long
as it takes for that industry to become established and to
bring its costs in line with those of major international
competitors. As soon as the industry becomes competitive
internationally, it is no longer in either the world's
(nor even particularly in the nation's) best interest for
the tariffs to be retained. What happens all too often,
however, is that the industry comes to consider the

protective tariffs to be something of a natural right, and hence the industry seeks to extend (and usually succeeds in extending) the life of the tariffs considerably past a reasonable deadline. Hence, the second point--high, protective tariffs should have associated with them a fixed and definite schedule of reduction.

It was noted that high, protective tariffs (or other barriers to trade) are not exclusively associated with the infant industries of developing nations. Cases can be found in most of the developing nations of industries that are not infant yet receive protection, and, more significantly perhaps, numerous cases of protection can be found among the industrialized nations. The contention of U.S. labor is that the net result is a discrimination against U.S. exports. However, it was pointed out that somewhat embarrassingly, for certain traded goods, it is the United States that has erected the high barriers to trade. Labor representatives maintain that in spite of this, on the balance, the barriers to importation into the United States are low in comparison to most other nations. Labor stands in favor of retaliation against what it considers to be unfair tariff practices by U.S. trading partners.

Generally, the workshop was against tariff retaliation. It was felt that retaliation could lead to a "tariff war," wherein all nations would begin to raise their tariffs in retaliation against one another to the net loss of all. Furthermore, were the U.S. to begin to impose retributive tariffs, it would either have to abandon many of its bilateral most-favored-nation treaties with its trading partners or to impose uniformly high tariffs on protected goods irrespective of the nation of origin. The specter posed was that of a world in which all nations were manipulating tariffs to gain short-term national advantage but in the long run forcing themselves toward autarkic positions. Such a situation, in the view of some analysts, was a major contributing factor to the Great Depression of the 1930's.[15]

Participants' response to the issue of whether or not the government (or organized labor) should have a voice in whether or not a firm closes a plant was mixed. It was pointed out that the federal government already has some regulatory power in this domain. Most notably, if a union or a group of workers believe that a firm closes down a plant and shifts production to another location (domestic or overseas) in order to avoid collective bargaining or to avoid a union organization effort, the union or workers may file a petition with the National Labor Relations Board (NLRB), which then does have the power to force the reopening of the plant should it find the complaints to be substantially correct. Labor representatives noted that while in fact this might be true, it is extremely difficult to prove that a firm shifts production overseas specifically to avoid collective bargaining.

Most participants in the workshop expressed a desire that the government not be given any further power to regulate or control the location of economic activity beyond that which it already has. The feeling ran strongly that the efforts of the federal government to regulate industry often lead to reduced efficiency with little or no net social gain. This feeling against federal regulation of any sort largely offset sympathy with the labor position.

3. THE LABOR POSITION AND THE INTERNATIONAL ECONOMIC POSITION OF THE UNITED STATES

One participant in the workshop (not a representative of organized labor) noted that however much one might agree or disagree with labor's concerns over the nation's international economic policies, the "voice" of labor is necessarily and rightfully a strong one in a democratic nation such as the United States. If, in a democracy, a large constituency of the voting population feels that the international economic policies (or lack thereof) of the government are damaging to the self interests of that constituency, there will result significant pressures to change those policies. This will occur no matter what the intrinsic merit of those policies might be.

At this time, this participant argued, U.S. labor clearly is a constituency which believes that it is aggrieved by the conduct of U.S. firms and the U.S. government in the international economy. Attempting to determine why there should be disaffection within the organized labor movement, this participant argued that the principal reason was that decisions affecting the welfare of U.S. labor are made on an almost daily basis but that labor itself is not consulted and that the effects of the decisions upon labor's interest are not taken into account. Such decisions are made both within large U.S. corporations and within the U.S. government. Given the inability of labor even to have an input into such decisions, it is to be expected that labor reacts against the decisions. Even if it could be assumed that such decisions ultimately do reflect labor's best interests (and there is little evidence to support such an assumption), labor might be expected to view the outcomes with a certain amount of jaundice. If one has no participation in decisions affecting one's welfare, one cannot help but be somewhat suspicious of those who do make the decisions.

To this participant, the key to accommodating labor's disaffection was therefore to create means and mechanisms by which U.S. labor could have some direct input into such key decisions as, for example, a firm's decision to invest abroad or the U.S. government's recent decision to encourage U.S. firms to invest in East European nations. Without such a voice, it was argued, U.S. labor could hardly be expected either to moderate its present position or to refrain from seeking drastic changes in U.S. international economic policy via the legislative process. One possibility would be for labor to participate directly in the decision-making process at the level of the firm. It was pointed out that in some industrialized nations, such as Sweden and West Germany, labor representatives sit on both corporate boards of directors and governmental high commissions. In Japan there is growing pressure for such representation.

A number of participants disagreed with the view that direct "worker participation" in company management would be desirable. Some felt that labor's role in the decision-making process is already considerably stronger than was suggested. Although labor representatives do not sit on the boards of directors of U.S. corporations, U.S. labor itself generally does not favor codetermination of this sort. On Capitol Hill and in certain federal agencies, however, the influence of labor is quite strong, and labor has often forced the passage of laws affecting business decision making. By means of the collective bargaining process, as enforced through the National Labor Relations Act, and the right to strike, labor can act to affect decision making by large corporations.

It was pointed out that if U.S. corporations were to expand the role of labor in their internal decision-making processes, it would be less than just to limit such a role to U.S. labor only. Many of America's large corporations are, of course, multinational and employ large numbers of non-U.S. workers. If a corporation has extensive international operations, it could be argued that labor representation should also be international.

Despite objections, there was a significant amount of support among many (but not all) of the workshop participants for greater labor representation in the making of decisions that affect labor interests, including support for greater participation in general trade negotiations and bilateral agreements. Support for such representation came from businessmen as well as other workshop participants. There was, however, little agreement over how to give effect to such a principle. It was recommended that the matter should be studied further.

One participant noted that, in his opinion, the most critical labor problem of the future would not be a shortage of jobs but a shortage of workers, especially in the higher-skilled job categories. This participant's opinion was that present labor efforts to protect jobs in low-skilled categories might have the effect of inhibiting the flow of resources into growing industries and thus locking workers into these job categories. It was further noted that a large percentage of persons who face losing employment in unskilled job categories (or are presently unemployed) are young, untrained, and often of minority backgrounds. It was recommended that programs are urgently needed to upgrade the skill levels of such persons and that rather than fighting to save unskilled jobs, the labor movement would be better advised to fight to have such programs created.

4. THE NEED FOR EXPANDED ADJUSTMENT ASSISTANCE

Whatever might be the nature of future demand for labor, the most immediate concern of organized labor leaders at the moment is that of unemployment. As previously noted in this chapter, fear of job loss is a major determinant of the protectionist sentiment prevalent among American labor today. It was believed by a majority of the workshop that unless measures are taken to reduce the trauma associated with job loss, protectionist sentiment among U.S. workers would remain high.

To dispel labor's anxieties, two ingredients are necessary. First, there must be growth in the economy, growth that will generate new job opportunities. Second, there must be some means or mechanism by which workers whose jobs are displaced because of relocation of economic activity can avail themselves of new job opportunities. Federally sponsored adjustment assistance is an important aspect of this second ingredient.

It was the opinion of a majority of the workshop that federally sponsored adjustment assistance programs must be enlarged and improved. Several specific areas for improvement are touched upon here.

Present TRA programs, as was noted in the second section of this chapter, provide adjustment assistance only for workers whose jobs are lost directly as the result of imports. It was felt that this requirement is excessively restrictive because many workers whose jobs have been lost to imports indirectly or to relocation of an American firm's activities abroad are not eligible for adjustment assistance. It was recommended that adjustment assistance programs should be implemented for all workers whose jobs are lost because of relocation of economic activities and not just those whose jobs are "lost" directly to imports.

It was noted that job retraining is conducted under the present TRA program, but it was felt that the extent of such retraining is inadequate. Expansion of programs to find new job opportunities for displaced workers and to train workers to hold new jobs was recommended. Because it might be necessary for a displaced worker to relocate in order to find new employment, a program of relocation assistance should be created.

Many workers whose jobs are lost because of relocation of economic activities are persons who are older and who might experience a considerable degree of difficulty in finding new employment or, if it could be found, in adjusting to it. In some such cases, it might be arguable that it would be better to provide such a person with a retirement income than to attempt to place that person in a new job. It was therefore felt that the possibility of creating an early retirement program for displaced workers for whom there do not exist alternative employment opportunities ought to be studied.

Such a program, it was realized, would provide a worker whose job was threatened by relocation of economic activity with economic security, albeit not with the personal satisfaction of holding employment. Therefore, any such early retirement program should be considered only in cases where a displaced worker was patently unable to find alternative employment.

It was recommended that both the maximum amount and the maximum duration of adjustment assistance payments to displaced workers be increased. Because the finding of new job opportunities and the learning of new skills is a costly and time-consuming process, it was felt that the level and extent of present TRA payments are inadequate.

It was recognized that implementation of the above recommendations would be costly and that the cost would be borne by the federal government. It was, however, also recognized that the alternative to these recommendations might be mounting pressure for restrictions on international trade and investment. Such restrictions, if implemented, could be much more costly to U.S. society than would be an expanded adjustment assistance program. Unrestricted trade and investment lead to increased efficiency in the international allocation of resources and hence to higher aggregate income for the United States as a whole. For the costs of this trade and investment to fall largely on specific sectors of U.S. labor while the benefits are captured elsewhere, it was felt, is both unjust and politically infeasbile.

NOTES

1. See, for example, William W. Winpisinger (General Vice-President, International Association of Machinists and Aerospace Workers), "Remarks at the U.S. Department of State National Meeting on Science and Technology," Mimeo (Washington, D.C.: International Association of Machinists and Aerospace Workers, 1976).
2. If labor means transfer of technology by means of foreign direct investment, the statement is not particularly accurate. U.S. textile, leather, and shoemaking firms generally have not, at any time in the twentieth century at least, made significant investments abroad. Throughout the twentieth century, and especially in the post-World War II era, the industries in which U.S. firms have made significant overseas investments in manufacturing have been foodstuffs, petroleum refining, chemicals, pharmaceuticals, rubber products, nonferrous metals, automobiles, electrical machinery (including electronics), and nonelectrical machinery. Electronics assembly has been one of the most recent of industries in which foreign direct investment has occurred. See Mira Wilkins, *The Emergence of Multinational Enterprise* (Cambridge, Mass.: Harvard University Press, 1970) and *The Maturing of Multinational Enterprise* (Cambridge, Mass.: Harvard University Press, 1974).

Labor leaders argue, however, that technology transfer in lower-skilled industries has been accomplished by means other than foreign direct investment. The example given is that, in the textile and clothing industries, U.S. buyers abroad specify quality, production methodology, and packaging in the placement of orders. This is seen by union leaders as being a "giveaway" of U.S. production technology.
3. Of course, there are other provisions in the tax code that favor domestic investment, most notably the investment

tax credit. See the discussion in the second section of this chapter.

4. It was noted that the imbalances are not all one way. Tariff and nontariff barriers to trade are very high with respect to importation into the United States of chemicals and pharmaceuticals. On some products, most notably cotton textiles and specialty steels, there exist U.S. import quotas.

5. See "Brazil: The Aircraft Industry Irks U.S. Competitors," *Business Week*, October 11, 1976.

6. The assumption of labor unions is that both increased importation resulting from changed terms of trade and foreign direct investment by U.S. firms cause a loss of jobs, in the short run at least, in the U.S. market. This assumption is in fact by no means clear. A debate has been pursued over the employment effects of trade and investment. An articulate summary of labor's point of view on this issue is given by Nathan Goldfinger, "A Labor View of Foreign Investment and Trade Issues," in R. E. Baldwin and J. D. Richardson, editors, *International Trade and Finance* (Boston: Little, Brown, and Company, 1974). For a counterview, see Robert B. Stobaugh, *Nine Investments Abroad and their Impact at Home* (Cambridge, Mass.: Harvard University Press, 1976).

7. An analysis of this type discussed by the workshop was Peggy Musgrave's report to the Subcommittee on Multinational Corporations of the U.S. Senate. Using an econometric model embodying the assumptions stated in the text, Musgrave estimated that had all direct investments made abroad by U.S. corporations been made in the United States instead, labor's income would have been from 6.7 percent to 15.5 percent higher in 1968 than was actually the case. The workshop considered Musgrave's analysis to be somewhat simplistic and, in particular, noted that many of her assumptions were biased in the direction of showing an income loss to labor. Even if the model contained no such bias, it was further noted, the income shifts reported by Musgrave were not extremely great.

A similar analysis, but using a more complex model, was reported by Lester C. Thurow and Halbert White. Using this model, it was shown that "optimal" restrictions on capital outflow for the United States would improve labor income by some 10 percent.

See Peggy B. Musgrave, *Direct Investment Abroad and the Multinationals: Effects on the United States Economy*, Report to U.S. Senate Subcommittee on Foreign Relations (Washington, D.C.: U.S. Government Printing Office, August

1975). See also L. C. Thurow and Halbert White, "Optimum Trade Restrictions and their Consequences," *Econometrica*, July 1976.

8. G. R. Neuman, "An Evaluation of the Trade Adjustment Assistance Program," unpublished report to the U.S. Department of Labor, Washington, D.C., 1976. The authors indicate that the sample of workers who were laid off their jobs because of imports may have been biased in the direction of including a disproportionately high number of older and less-educated workers.

9. The base wage is set equal to the highest average weekly wage (including wages paid from a second job) paid during any of the four quarters prior to layoff. Maximum weekly payments, however, are $190.00. A qualifying worker can receive up to 52 weeks of TRA benefits, which can be applied retroactively up to 1 year from the date of the filing of the petition and can be carried forward for up to 2 years from the date of filing.

10. These criteria are somewhat less stringent than those of the 1962 act, wherein for an unemployed worker to be eligible for TRA, it was necessary for certification by the U.S. Tariff Commission that a worker lost his job due to imports resulting from concessions granted under trade agreements.

11. See George R. Neuman, "The Direct Labor Market Effects of the Trade Adjustment Assistance Program: The Evidence from the T.A.A. Survey," unpublished paper for the U.S. Department of Labor, Washington, D.C., 1977.

12. A Labor Department official contacted by the reporter indicated that a worker could be eligible for TRA if the worker's employer closed the domestic plant at which he was employed and imported from the overseas plant to serve the U.S. market formerly served from the domestic plant. If, however, the overseas plant produced primarily for local overseas markets, the official agreed that the worker probably would not qualify for TRA.

13. While a consensus was claimed, there was not unanimous agreement among the nonlabor participants with respect to these points. Several participants did believe that there might be arguments for direct controls in certain cases. (These cases are taken up in the next two chapters.)

It was generally felt that the main function of a readjustment assistance program is to increase labor mobility, but whether mobility could ever be achieved without forcing some workers to assume a personal welfare loss was doubted by several participants. It was agreed, however, that the best hope for workers displaced by trade or investment is in the direction of upgrading their skill levels. It was noted that the vast majority of workers who lose jobs because of imports are in the low-skill labor categories and that many of these workers could be reabsorbed into the economy at higher skill levels were they to be given the prerequisite training.

14. It is agreed, however, by most individuals who have studied these problems that elimination of the tax deferral without elimination of the foreign tax credit would not alter the U.S. foreign direct investment position by a large amount. Elimination of the tax credit, either with or without elimination of the deferral, would have a much larger effect on the U.S. foreign direct investment position, but how large an effect is difficult to determine. For (differing) estimates of the effects (and conflicting opinions on what actions would be appropriate), see Peggy Musgrave, "Tax Preferences to Foreign Investment," in *Economics of Federal Subsidy Programs, Part 2--International Subsidies*, Joint Economic Committee, 92d Congress, 2d Session 176 (1972); Peggy Musgrave, *Direct Investment Abroad and the Multinationals*; U.S. Tariff Commission, *Report to the Committee on Finance of the United States Senate on Implications of Multinational Firms for World Trade and Investment and for U.S. Trade and Labor* (February, 1973); and Thomas Horst, "American Taxation of Multinational Corporations," Mimeo (Medford, Mass.: Tufts University, September 1975).

15. See C. P. Kindleberger, *The World in Depression* (Berkeley, Calif.: University of California Press, 1970), chap. 14.

5 U.S. TRADE AND TECHNOLOGY TRANSFER TO THE SOVIET UNION AND THE EASTERN EUROPEAN NATIONS

1. THE MAJOR ISSUES

The major issue with respect to U.S. trade with the Soviet Union, as identified by the workshop, is whether and to what extent the Soviet Union ought to be treated as a trading partner on the same basis as any other nation. Several factors suggest that the Soviet Union is not simply an ordinary trading partner. It is a very large nation, richly endowed with natural resources and having the world's second largest national economy. Soviet external trade is conducted through monolithic state buying and selling agencies. Analysts note that the Soviet Union therefore can act as a monopolist and monopsonist in international commerce to improve her terms of trade.[1] The Soviet Union is a rival to the United States in terms of military power and world influence, and it is an objective of the United States that the Soviet Union not receive by means of trade any technology that might enable the Soviet military significantly to increase its capabilities relative to those of the United States.

It was agreed by most workshop participants that mutual tangible gains can be realized by both the United States and the Soviet Union from trade. Whether or not the net gains accrue equally to both partners was contested, with a number of participants believing that the gains were skewed to the advantage of the Soviet Union.

Primary among Soviet needs for trade are two classes of goods: grains to supplement harvests that fall short of target and advanced capital goods to bolster productivity. (See the third section of this chapter.) These needs are real, tangible, and immediate.

The potential gains to the United States from trade with the Soviet Union are not so immediately important to the

U.S. economy. Many of the alleged gains, in fact, are quite intangible. Among the possible long-term gains to the United States from trade with the Soviet Union are the following, as discussed by the workshop:

1. An increase in the economic interdependence of the West and the Soviet Union might reduce the probability of war occurring between the United States and the Soviet Union.

Most participants believed that this was in fact the case. It was pointed out, however, that historically there is little relationship between increased flows of trade between nations and reduced propensity of these nations to war with one another. For example, trade between Germany and most of the Allied nations increased substantially between 1900 and the outbreak of World War I. In this case, increased trade apparently did not deter the outbreak of the war.

2. Trade might serve to raise the standard of living (and hence the aspirations) of the Soviet people. The net result might be an internal reallocation of resources within the Soviet Union from the military to the civilian sector.

This proposition was supported by many workshop participants but was seriously disputed by some. The counterargument was that Russia during the late nineteenth century and again during the 1930's acted to expand trade with the West greatly and to accelerate efforts to absorb Western technology.[2] At neither time was there a significant perceptible change in Russian priorities with respect to overall allocation of resources. Russia, it was pointed out, has historically demonstrated an ability to absorb Western goods, technology, and capital without absorbing Western attitudes. Furthermore, the Russian state, both under the Czars and under the Communists, has historically demonstrated an ability to force the Russian people to accept a reduced material standard of living in order to allocate resources to meet national goals.

3. Trade may enable the United States to exact a *quid pro quo* in bargaining with the Soviet Union on nontrade matters (such as the Strategic Arms Limitation Talks).

This assumes that the Soviet Union needs trade with the United States more than the United States needs trade with the Soviet Union--something which many workshop participants believed to be the case. It must be noted, however, that what the Soviet Union seeks and needs is trade with the West and not necessarily trade with the United States.

To the extent that Western nations other than the United States are able and willing to sell goods or technologies to the Soviet Union on terms more favorable than those granted by the United States, the bargaining power of the United States is clearly reduced.

4. Trade with the Soviet Union might lead to an economic gain for the United States (or for the West).

This is, of course, just the standard gains-from-trade argument. In order for there to be such a gain, it is necessary that the trading partners of the Soviet Union collectively be able to import goods from the Soviet Union on terms more favorable than those on which the goods could be obtained elsewhere. This presupposes that the Soviet Union has goods to export that are sought in the West and that the Soviet Union is able and willing to offer these for sale at favorable prices. Both of these suppositions are open to some question. In recent years the Soviet Union has been running a major trade deficit with the West and has been forced to finance this deficit with short-term and medium-term credit.[3] (See Table 8.) While credit can serve in the short-term to finance disequilibria brought about by a growth in East-West trade, in the longer term imports by the Soviet Union can be increased only if there is a corresponding increase in exports. Inability to export has previously been disruptive to Soviet-Western commercial activities. During the 1930's, for example, the

TABLE 8 Soviet Union Hard Currency Trade Balance, 1970-1975 (U.S. $ millions)

	1970	1971	1972	1973	1974	1975
Exports	2,197	2,652	2,815	4,818	7,630	7,800
Imports	2,711	2,955	4,171	6,566	8,541	14,081
Net imports	514	303	1,356	1,748	912	6,281
Estimated hard currency debt at year-end[a]	1,722	2,029	2,608	3,641	4,461	7,489

[a]Medium and long term.

SOURCE: *Soviet Economy in a New Perspective* (papers presented to the Joint Economic Committee, 94th Congress, 2d Session, October 1976), pp. 728, 738.

Soviets found themselves in a position of not being able to pay for imports of machinery and Western technical assistance, and as a consequence between 1932 and 1934 Soviet purchases in the West dropped sharply. For the next 25 or so years (excluding the war years) the Soviet economy functioned on an autarkic basis. It has been argued that a return to autarky by the Soviet Union is not out of the realm of possibility in the future and that prevention of such a return requires that Soviet exportation efforts be sharply increased.[4]

At any rate, it is not absolutely clear that the West is deriving much, if any, purely economic benefit from trade with the Soviet Union at this time, given the magnitudes of Soviet trade deficits. Continued trade deficits by the Soviet Union could conceivably result in defaults on credit, which, although highly unlikely, would certainly be damaging to the West. More likely, continued deficits would eventually force the Soviet Union to curtail trade with the West, which would be highly disruptive to those firms that do substantial business with the Soviet Union.[5]

The critical question is whether the benefits to the United States of trade with the Soviet Union, both tangible and intangible, are greater than the total costs. The consensus of the workshop was that an assessment of the net benefit of trade with the Soviet Union is predicated upon some understanding of the long-term intentions of that nation. In the extreme, if the Soviet Union were to be preparing to wage early war with the West, it would be in the West's best interests to discontinue trade immediately. Fortunately, war is not perceived to be a likely event. In the opposite extreme, were it the intention of the Soviet Union to evolve into a free market economy and to terminate its political rivalry with the West, it might well be appropriate for the United States and other Western nations to reduce restrictions in trade with the Soviet Union to a degree comparable to that prevailing among the OECD nations.

In the opinion of some experts, the Soviet Union seeks to pursue a policy of what has been termed "competitive detente."[6] To the extent that this means that the Soviet Union intends to foster long-term trading arrangements with the West, it is probably in the interests of the West to reciprocate but with caution. The reasons for caution are taken up in Section 4 of this chapter.

The major problem with all of this, of course, is that no one, perhaps not even the Soviet leaders themselves, is exactly sure of what Soviet intentions are. The best that can be said is that Soviet intentions remain somewhat obscure, and given this obscurity, the policy alternatives open to the United States must remain somewhat flexible.

2. TRADE WITH THE EAST EUROPEAN (COMECON)* NATIONS

Many of the arguments for U.S. trade with the Soviet Union can be applied to the case of U.S. trade with the Eastern European nations. Of special significance is the question of economic interdependence among the COMECON nations. It has been noted that U.S. official policy during the 1950's and early 1960's of discouraging East-West trade unavoidably resulted in a high degree of economic interdependence between the Soviet Union and other COMECON nations. By blocking any effective commerce between East Europe and the Western industrialized states, the West left East Europe with no choice but to obtain raw materials, capital, and industrial technology from the Soviets.

It is now generally recognized that the economic dependence of the East European nations upon the Soviet Union significantly enhances the ability of the Soviet Union to exert influence over these nations. A very legitimate objective of expansion of economic links between these nations and the West is therefore to attempt to reduce economic dependence of Eastern Europe upon the Soviet Union. While it would doubtlessly be naive to assume that expanded commercial relations between Eastern Europe and the West could lead to an ending of Soviet domination in that area of Europe, expansion of commercial relations with the West does at least give Eastern Europe an alternative to dependence upon the Soviet Union. To the extent that this alternative can be pursued by Eastern Europe, at least the possibility exists of Eastern European nations moving away from a position of total alignment with the Soviet Union toward a more neutral position.

While most workshop participants agreed with this last proposition, the reminder was made that the political cohesiveness among the COMECON nations is not wholly or even largely a result of economic interdependence. Rather, cohesiveness results from a common ideology among these nations, and, more importantly, Soviet military dominance in Eastern Europe.

---

*COMECON is Council for Mutual Economic Assistance--a trade organization of Eastern European (socialist) countries.

3. THE BASIS FOR U.S.-SOVIET UNION TRADE: THE SOVIET POINT OF VIEW

It was noted by the workshop participants that the major expansion in commercial relations between the Soviet Union and the West that has occurred in the past decade or so has come about largely through initiatives of the Soviet Union. It was noted that throughout most of the post-World War II era the Soviet Union has been a relatively autarkic economy, trade having been largely limited to its Eastern European satellites and, to a much lesser extent, to a handful of developing nations. Because of the large size of the Soviet economy and the rich endowment of the Soviet Union with natural resources, she has been able both to be autarkic and to maintain a respectable economic growth rate for long periods of time.

Two sets of weaknesses in the Soviet economy have contributed in large measure to efforts by the Soviet Union to increase economic exchange with the West. The first has been the internal Soviet agricultural problem. The productivity of the agricultural sector has been one of the slowest-growing aspects of the Soviet economy since at least the 1930's. Consequently, despite the fact that the Soviet Union employs an unusually large proportion of its labor force in agricultural undertakings (see Table 9), the Soviet economy has often faced an undersupply of agricultural products in recent years. Thus, the Soviet Union has been forced periodically to import large quantities of grain and other foodstuffs during years of undersupply, particularly in 1963, 1972, and 1975 (see Table 10). Most experts feel that improvement of productivity is a high-priority item for Soviet economic policy makers and that there has been little change in the Soviet Union's agricultural policy for over a decade. The policy of the past decade has been both to devote considerable resources to the agricultural sector and to upgrade the diet of the average Russian citizen. As a consequence, there has been an apparent long-term improvement in the productivity of Soviet grain farms, but increased demand for grain for the production of meat has outstripped productivity gains. Bad harvests—caused by the weather rather than by any decreased commitment to the agricultural sector—in 1972 and 1975 have exacerbated the Soviet Union's problems.

During the next 10 years, it is safe to expect that the Soviets will continue to make heavy capital investment in the agricultural sector, particularly to improve the productivity of grain farming. However, it is also likely

TABLE 9 Urban and Rural Populations of the Soviet Union Compared with the United States and Western Europe, 1970

	Population (thousands)	Percent
Soviet Union		
Rural	104,044	42.9
Urban	138,568	57.1
Total	242,612	100.0
United States		
Rural	53,297	25.9
Urban	152,209	74.1
Total	205,506	100.0
Western Europe[a]		
Rural	38,055	25.6
Urban	110,564	74.4
Total	148,619	100.0

[a]Austria, Belgium, France, W. Germany, Liechtenstein, Luxembourg, Monaco, Netherlands, and Switzerland.

SOURCE: United Nations, *Statistical Yearbook, 1972*.

TABLE 10 Composition of Soviet Union Hard Currency Imports, 1974 and 1975 (U.S. $ millions)

	1974		1975	
	Value	Percent	Value	Percent
Machinery and equipment	2,333	27.3	4,553	32.3
Metals	2,628	30.8	4,097	29.1
Chemicals, plastics, rubber	983	11.5	936	6.7
Manufactured consumer goods	261	3.1	428	3.0
Foodstuffs	1,082	12.7	3,203	22.7
Grain	523	6.1	2,298	16.3
Other	1,254	14.6	864	6.1
TOTAL	8,541	100.0	14,081	100.0

SOURCE: *Soviet Economy in a New Perspective* (papers presented to the Joint Economic Committee, 94th Congress, 2d Session, October 1976), p. 738.

that demand for meat and other high-grade foodstuffs will continue to grow, causing increasing demands for grain. Thus, in 1975 the Soviet Union signed an agreement with Western wheat producers that commits the Soviet Union to import 6 million tons of grain per annum for five years.[7]

The second set of weaknesses in the Soviet economy has been low factor productivity in a number of subsectors of the manufacturing sector. The causes of this low factor productivity are only partially understood, but a number of observations can be advanced.

The Soviet Union in a number of industries distinctly lags behind the United States and other Western nations in the development and application of technology. This has occurred in spite of the fact that OECD data suggest that the Soviet Union spends a higher percentage of its national product on R&D than does any other nation (see Figure 1) and employs a higher percentage of its population as scientists and engineers than does any other nation (see Figure 2). Three reasons stand out for the technological backwardness of certain Soviet industries:

First, Soviet efforts in R&D have been heavily oriented toward military and aerospace activities, and there has been very little "spillover" from these into the civilian economy. The propensity of the Russian military to enforce secrecy has doubtlessly served to minimize any spillover that might otherwise have occurred.

Second, the classical Soviet growth strategy has placed very heavy emphasis upon maximizing the rate of addition of labor and capital to the industrial sector, with relatively little emphasis upon technological modernization of most industries. This strategy has caused the Soviet Union to generate both a very high rate of new capital investment and a rapid rate of transfer of workers from the farm to industry. As a result, the Soviet economy has achieved a moderately high rate of growth during the two decades spanning from approximately 1947 to 1967. This rate of growth was accomplished without widespread technological innovation within the Soviet economy because the economy was capital short and because resources were poorly allocated. In the past 10 years or so, however, Soviet economic planners have noted diminishing returns from new capital investment and have further noted the prospect of a decline in the rate of growth of the industrial labor force. Consequently, Soviet planners increasingly have turned to technological progress as a means of promotion of economic growth.[8]

Third, the Soviet economy is one in which the allocation

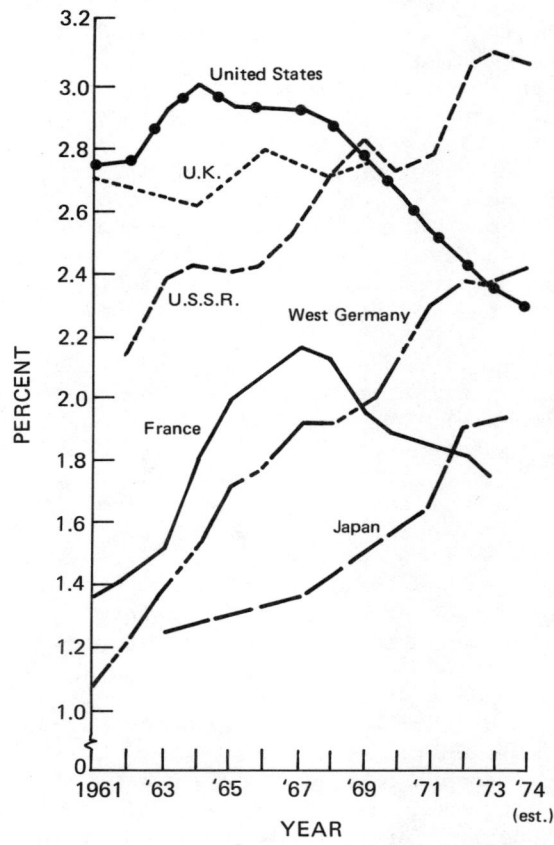

SOURCE: Organisation for Economic Co-operation and Development; individual country sources; U.S.S.R. estimates by Robert W. Campbell, Indiana University.

FIGURE 1 R&D expenditures as a percent of GNP, by country, 1961-1974. From National Science Board, *Science Indicators 1974* (Washington, D.C.: National Science Foundation, 1974).

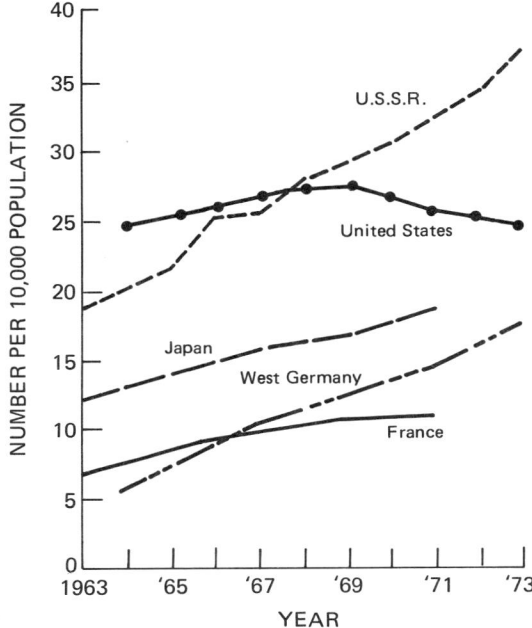

SOURCE: Organisation for Economic Co-operation and Development; individual country sources; U.S.S.R. estimates by Robert W. Campbell, Indiana University.

FIGURE 2 Scientists and engineers engaged in R&D per 10,000 population, by country, 1963-1973. Includes all scientists and engineers (full-time equivalent basis). Data for the United Kingdom are not available. From National Science Board, *Science Indicators 1974* (Washington, D.C.: National Science Foundation, 1974).

of resources is determined by central planning rather than by the competitive marketplace. In the opinion of some analysts, the result is a suppression of innovation in many sectors in the Soviet economy causing these sectors to lag behind their Western counterparts in the application of factor savings technology.

There is little doubt in anyone's mind that the Soviet Union could, should it choose to do so, upgrade its industrial technologies without major assistance from the West.

The general belief is, however, that for the Soviet Union to do so would be extremely costly and might necessitate the transfer of scientists and engineers from the military/aerospace sectors to the civilian economic sectors. Hence, for reasons of both efficiency and national security, the Soviets have been turning toward the West to obtain commercial technology. This is reflected in the large purchases by the Soviets of Western machinery in recent years (see Table 10) and by the increasing willingness of the Soviets to enter into technical assistance agreements with Western corporations. The Soviet intention apparently is to use Western technology to improve Soviet civilian economic performance.[9]

The following question was raised: If Soviet purchases of technology allow internal R&D efforts to continue to be concentrated in the military and aerospace endeavors, is there a case to be made for restricting the sale of civilian technology to the Soviet Union on military grounds? Possibly, by refusing to sell technology to the Soviet Union in any form, the West could effectively force the Soviets to reallocate resources from "guns" to "butter."

Two objections were registered to this type of reasoning. The first was that it is not clear that even if a successful embargo on the sale of technology to the Soviet Union could be achieved, it would necessarily lead to any reduction in Soviet military expenditure. More likely, it was felt by the workshop, the Soviets would content themselves with a slower pace of modernization of the nonmilitary industrial sector and would continue to allocate as much of their total resources to the military as they would have even if no embargo had been instituted.[10]

The second, and perhaps more compelling, objection was that an embargo against the sale of technology to the Soviet Union would generally not be practicable. The technologies most needed by the Soviet Union to improve factor productivity are mostly ones that are quite mature and standardized and that are available from numerous sources among the world's nations. It was felt by workshop participants, some of whom themselves had had the opportunity to engage in commercial activities with the Soviet Union, that unilateral U.S. actions to block the sale of industrial technology to the Soviet Union usually result simply in the purchase by the Soviets of the technology from some other Western nation. Any effort to blockade sales of technology to the Soviet Union would therefore have to be pursued multilaterally by all industrialized Western nations, and the probability of getting all potential sellers of commercial technology to agree to a multilateral blockade, it was felt, would be next to nil.

TABLE 11 Composition of Soviet Union Hard Currency Exports, 1974 and 1975 (U.S. $ millions)

	1974		1975	
	Value	Percent	Value	Percent
Fuels	2,905	38.9	3,763	48.3
Metals	582	7.6	328	4.2
Wood and wood products	1,032	13.5	699	9.0
Cotton fiber	357	4.7	289	3.7
Furs and pelts	71	0.9	65	0.8
Other (includes manufactured goods)	1,434	18.8	1,645	21.1
Unspecified[a]	1,247	16.3	1,010	12.9
TOTAL	7,630	100.0	7,800	100.0

[a]Composed mostly of diamonds, precious metals, and nickel.

SOURCE: *Soviet Economy in a New Perspective* (papers presented to the Joint Economic Committee, 94th Congress, 2d Session, October 1976), p. 738.

A major problem with trade with the Soviet Union, as was expressed by the workshop, is getting something in return. Soviet exports to the West have largely consisted of fuels and other raw materials, and only a small percentage of the total has consisted of manufactured goods. (See Table 11.) Expanded trade with the Soviet Union, to be of any economic advantage to the West, must (as was stated in the previous section) involve the willingness and capacity of the Soviets to sell to the West desired goods on favorable terms on a continuing basis. Also, reservations were expressed about the growth of barter agreements between the Western nations and COMECON nations, which have the effect of reserving sections of Western markets for COMECON-produced goods.

4. CONSTRAINTS ON THE DESIRABILITY OF EXPANSION OF EAST-WEST TRADE

The workshop identified several factors that might act as constraints on the desirability of expanded East-West trade. Among these factors are differences in institutional

structure between the economies of the East and those of
the West, the pricing of industrial technology sold by the
West to the Soviet Union, and the possible adverse effects
of expanded East-West trade upon the Western trading and
payments system.

Differences in Institutional Structure between the
Eastern and Western Economies

It goes virtually without saying that the economy of the
Soviet Union (as well as those of the other COMECON nations)
is organized very differently from the economies of the
Western industrialized states. Economic institutions
within the Soviet Union are state owned and are subject
to centralized economic planning. The international trade
of the Soviet Union is conducted through monolithic state
buying and selling agencies. The economic institutions
of the Western states are largely privately owned and are
subject either to no centralized state planning or to very
limited state planning. Western nations' trade is largely
conducted through a large number of privately owned institutions.
 The point was raised briefly earlier in this chapter
that the conduct of trade by the Soviet Union through monolithic state trading agencies might enable the Soviet Union
to establish terms of trade that are highly favorable to
her interests. For example, if the Soviet Union were to
export a product for which she was a unique or dominant
supplier, she could act as a monopolistic price setter in
world markets for this product, setting a price to realize
maximum revenue from exportation. In fact, however, the
bulk of Soviet exports to the West in recent years have been
standard commodities for which there have been alternative
sources of supply and for which the Soviet Union has not
been a dominant source. Thus, the Soviet Union, as an
exporter, has largely behaved as an international price
taker rather than as a price setter. Furthermore, she has
usually been too small a factor in the market to affect
international price levels significantly. This, however,
might not always be the case. Especially if trade between
the Soviet Union and the West were to continue to expand,
the possibility exists that the Soviet Union could become
a price setter in world markets for certain classes of
exports.
 On the import side, all importation into the Soviet

Union is conducted through state trading agencies, which serve as monopsonistic buyers of foreign goods for the Soviet economy. Such a monopsonistic buyer, if it faces competitive sellers, holds the power to force the sellers to bid against each other for the privilege of making the sale. If the sellers do compete, the final transaction price will be equal to the marginal cost of the good.

Conceivably, then, the differences in institutional structure between the economy of the Soviet Union and those of the Western states might result in the Soviets being able to capture a rent on exports but never paying a rent on imports. The result would be that the terms of trade would be more favorable to the Soviet Union than would have been the case were her economy institutionally similar to that of the Western states.

Technology Transfer and the Pricing of Technology Sold to the Soviet Union: Private Yield Versus Social Yield

It has already been noted that one of the reasons why the Soviet Union desires economic interaction with the West is to obtain Western technology, primarily to improve factor productivity. Soviet acquisition of Western technology has been concentrated in relatively few industries, apparently ones in which Soviet planners have identified a particularly large gap between their own and Western performance; these industries are consumer products, chemicals, motor vehicles, gas turbines, oil and gas transport, electrical equipment, computers, and industries supplying the agricultural sector. Presumably, the price that the Soviet Union would be willing to pay for such technology is bounded by the opportunity cost of developing technology internally.

A question raised at the workshop was whether or not the price actually paid by the Soviet Union is even close to the price the Soviet Union might be willing to pay for this technology. Western firms calculate the minimum price at which they can sell a technology by determining the marginal cost of the technology and setting the price equal to this marginal cost. This price is likely to be below the opportunity cost of developing the technology to the Soviet Union. If the Soviet opportunity cost is greater than the price at which Western suppliers are willing to sell the technology, the difference in institutional structures between the Soviet Union and the Western states are such that the transaction price will be the lower figure. This would especially be the case if there were multiple sources of supply for the technology and if these sources were domiciled in more than one nation. Soviet buying agencies could take advantage of the competitive structure of world markets, wherein potential sellers bid against one another. Because the technology most sought by the Soviet Union is proven, factor savings technology, it is likely to be available from multiple suppliers.

The net result, it was argued at the workshop, can be detrimental to the West. The Soviet Union, politically a strong adversary of the West, can purchase from the West needed factor savings technology at a relatively modest cost to the Soviet economy. This technology makes a contribution to Soviet productivity, thereby reducing pressure on Soviet policy makers to build up the civilian sector of the economy. How great this contribution is, as has been noted, is a matter of some controversy.

It must be noted that not all workshop participants agreed with these conclusions. Several participants, especially those having backgrounds from private industry, felt that private firms are generally able to negotiate with the Soviet Union prices that approach the maximum price that the Soviet Union might be willing to pay for a given technology. To support this, it was noted that (1) the Soviets, rather than engaging firms in a competitive bidding process, often prefer to work with just one firm and (2) the owner of a proprietary technology is often the only party equipped to judge the price that reflects the economic value of the technology. Thus, it was argued, the seller proposes a price that he believes would be acceptable to the Soviet Union, and the transaction is completed at that price.

It was agreed that this might in fact be the case for a specialized proprietary technology for which there were but one or a very few suppliers. Few participants believed, however, that a private seller could negotiate in this manner if there were numerous alternative sources of a particular technology, wherein the forces of competition would set the price.

Possible Adverse Effects of Expanded East-West Trade upon the Western Trading and Payments System

The Western trading and payments system consists of a complex array of institutions, treaties, and understandings, the functioning of which is premised upon several basic principles. Among these, perhaps the most important is that trade is conducted by a multitude of private firms, each of which acts to optimize its own economic welfare. Thus, in the absence of governmental interference in trade, private firms within a national generally will choose to export a product if international prices exceed those that can be obtained in the domestic market. Likewise, products will be imported into a nation if domestic prices exceed international prices. Both domestic and international prices themselves for most goods are determined by the interaction of supply and demand (rather than by governmental determination). Trade thus proceeds until international supply and demand are balanced, at which point prices (exclusive of transportation, distribution, and tariff charges) are equalized internationally. To the extent that relative price differences internationally and domestically determine the amount and direction of a nation's trade, economists reason that there are benefits to trade that generally accrue to both that nation and its trading partners.

The Soviet Union, by contrast, does not necessarily conduct its trade on the basis of differences between its own domestic prices for traded goods and international prices. Furthermore, domestic prices within the Soviet Union are most often administered by the government rather than determined by supply and demand, and goods for which there is excess demand are rationed. Thus, for example, in 1973 the Soviet Union exported cars to the West and imported wheat from the West, despite the fact that the relative price of wheat to cars within the Soviet Union was much lower than it was anywhere in the West. Both the Soviet domestic price of cars and wheat are administratively determined, and the relative prices of these goods made little or no difference to the Soviet planners when it was decided what to export and what to import. The calculation rather was based upon the social priorities of the Soviet Union at the moment.

As long as Soviet trade with the West is relatively small in magnitude and the Soviets are international price takers, it makes little or no difference to the West how the Soviet Union decides what to import or what to export. Problems could arise, however, when and if the Soviet Union becomes a large factor in international trade or when barter deals reserve special domains for Soviet products in Western markets. One problem is that the Soviet presence could shift prices away from those that would prevail when international supply and demand are in balance. This could lead to an undesirable reallocation of resources internationally in response to the new international price levels that would be set, for practical purposes, by the Soviet Union. A second problem, one more severe because it compounds the first, is that the Soviet Union could abruptly change its trading patterns, causing international supply or demand to surge suddenly and international prices to rise or fall precipitously and unpredictably.

The classic example is the sudden Soviet entrance into world oil markets in 1926. In order to obtain Western currency, the Soviets sold large quantities of oil to the West at prices well below the then-prevailing international price levels. The result was chaos in the Western oil industry, and the Soviet action is generally considered to have been one of the stimulants that led to the formation of an international petroleum sellers' cartel in 1928.

These problems are both compounded and ameliorated somewhat by the institutional aspects of some trading arrangements that presently are being made with the Soviet Union. Bilateral barter agreements between the Soviet Union and Western states, for example, cause goods to be traded outside of the framework of the international marketplace. In effect, corners of the markets of the Western economies are reserved for Soviet exports. Reservation of a share of a market for one supplier is a discriminatory practice. Such discrimination is generally limited or prohibited under GATT and, consequently, the practice is not carried out among GATT nations. The Soviet Union, however, is not a party to GATT, and thus GATT regulations are not applicable to bilateral barter agreements between the Soviet Union and Western nations. It was felt by some participants that the existence of such bilateral barter agreements violated the spirit of GATT and even posed a threat in the long run to the viability of the GATT framework. Offsetting this somewhat is the consideration that barter agreements are generally long term in nature, and thus the risk that Soviet trade will occur in surges is reduced.

5. WHAT CONTROLS SHOULD THERE BE ON U.S. TRADE WITH THE SOVIET UNION?

To a very large extent, discussion in the workshop focused upon what controls, if any, the U.S. government should exercise over U.S.-Soviet commercial relations. Several intertwined themes emerged. One theme involved the complex problems associated with military considerations in appraising whether and how the government might intervene in the transfer of technology to the Soviet Union. This led to a discussion of the balance of costs and benefits and problems of effectiveness in the bureaucratic control of selected categories of technologies and goods and services, even in situations in which such controls in principle may be deemed desirable. These problems of bureaucratic control, sufficiently difficult on a national basis, are compounded by the international implications of attempts to limit trade in goods and services available from many different nations.

A second theme involved a question of whether the benefits and costs of transfers of technology between the United States and the Soviet Union could be meaningfully analyzed in the familiar terms of international trade theory. In the discussion, several subthemes emerged within this second theme. One subtheme involved a belief held by many in the group--as well as by a number of commentators in the literature--that the advantages of current trade between the United States and the Soviet Union were skewed in favor of the Soviet Union. Another subtheme related to the monopolistic and monopsonistic nature of the Soviet Union's state trading apparatus in relation to the private enterprises of the United States and other Western nations and Japan. A third subtheme related to the implications of the growing trade deficit of the Soviet Union. In its simplest form, these implications were discussed in terms of whether the debt could and would be paid. In a more complex form, these implications involved an inquiry into whether the Soviet Union in fact had available exportable goods and services through which payment could be made and whether it would in fact be prepared to export goods and technologies that the West might find it desirable to acquire. In an extreme form, this second theme raised questions as to whether the United States, either unilaterally or in cooperation with other Western industrialized states, should give consideration to the possibility of direct governmental participation in East-West trade.

The First Theme: Control of Militarily Sensitive Technology

Many participants felt that efforts by the U.S. government to control the transfer of militarily sensitive technology to the Soviet Union are cumbersome and in some cases counterproductive.[11] A number of agencies are involved in control procedures, and the priorities of these agencies are often perceived to be in conflict.

The primary problem with the present system as a whole, it was felt, is that it is exceedingly difficult to determine what criteria should be applied to proscribe items from sale to the Soviet Union. Present U.S. export control procedures give the U.S. government a veto power only over such sales, and thus the procedures are oriented toward determining what to veto and what not to veto. Examples abound over the inconsistency of the procedural process; permission to export a product being granted but permission to export spare parts for the same product being denied is a typical story. Such problems arise from the complexity of drawing up a taxonomy of goods to be proscribed from sale to the Soviet Union.

A commonly told story is that of the U.S. government denying to a U.S. firm a license to export a good to the Soviet Union on the grounds of national defense, only for the Soviets to purchase the good from some other Western state. Workshop participants were somewhat divided on the question of whether or not this represents a real problem. It was pointed out that if the good is on the COCOM* list, its sale to the Soviet Union would be proscribed by all COCOM nations. If, on the other hand, the good were to appear on the U.S. Commodity Control or Munitions Control lists (but not on the COCOM list), it would be possible for a U.S. firm to be denied a sale to the Soviet Union that could be made by a firm domiciled in a COCOM nation other than the United States. Thus the consensus of the workshop was that these lists are too long and too specific.

Two propositions received general support from the workshop participants. The first was that either the control lists should be shortened or that simplified criteria for granting licenses to ship items on the list (but not items embodying militarily sensitive technology) to the COMECON nations should be established. (However, it was noted that the Commodity Control List has already been shortened considerably in recent years.) The second was that the procedure for granting licenses to export to the Soviet Union (and other COMECON nations) be simplified. Several participants felt that the procedures of the Bureau of

---

*COCOM is Coordinating Committee of the NATO countries plus Japan minus Iceland, which deals with trade items having strategic implications.

East-West Trade for granting export licenses are slow and inefficient, and it was recognized that at least part of this problem stems from the fact that the bureau often must receive approval from other federal agencies before a license is issued. The hope was expressed that a single agency could be charged with sole responsibility for the granting of export licenses.

An alternative scheme for the control of transfer of militarily sensitive technologies was discussed.[12] The basic premises underlying this alternative scheme were the following: (1) There is no Western technology, military or otherwise, that the Soviet Union is incapable of duplicating internally; the only militarily relevant technological advantage that the West has over the Soviet Union is that the "state of the art" of Western technology may at any point in time be more advanced than that of the Soviets. (2) Therefore, the key to the control of militarily sensitive technology lies in "lead time," which here is defined as the time between the introduction of a technology of a certain level in the West and the introduction of a comparable technology in the Soviet Union. (3) Lead time is a function of the level and rate of diffusion of design and manufacturing know-how (as opposed to either basic scientific knowledge, which, it can be presumed, the Soviets have, or knowledge of the specific design of a particular item). The major proposals suggested by the scheme were as follows:

1. Three categories of export should receive primary emphasis in control efforts: arrays of design and manufacturing information; "keystone" manufacturing, inspection, and automatic test equipment; and products accompanied by sophisticated information on operation, application, or maintenance.

2. The more active the transfer relationship, the more effective the transfer mechanism. Therefore, transfer mechanisms to be controlled most tightly are turnkey factories, joint ventures, training in high technology, licenses involving extensive teaching, and certain types of processing equipment.

3. To protect U.S. lead time, permission to export should be denied if a technology represents a *revolutionary* advance to the receiving nation but could be approved if it represents only an *evolutionary* advance.[13]

The primary problem with these proposals, it was felt, is that to set up an administrative system to implement the proposals would necessarily involve the creation of an administering agency having broad powers to investigate and control U.S. industry over a very wide range of activities. Proponents of the proposals countered that the intent of the scheme was to reduce the amount of bureaucracy involved in East-West trade and to reduce the total governmental scrutiny of commercial transactions between U.S. private industry and the Soviet Union. It was questioned, however, whether governmental efforts to control "arrays of design and manufacturing information; 'keystone' manufacturing, inspection, and automatic test equipment (i.e., equipment of a unique kind without which advanced weaponry could not be manufactured); and products accompanied by sophisticated information on operation, application, or maintenance" could possibly be consistent with a reduction of total governmental scrutiny or bureaucracy.

A feeling did emerge, however, that establishing criteria for controlling the transfer of militarily sensitive technology to the Soviet Union based upon lead time considerations was a useful approach. Thus, it was felt that sale to the Soviet Union of a technology (or a good embodying a technology) should be proscribed only if the technology was one in which the United States held a significant lead time advantage. If no such advantage were to exist, the implication being that the technology is available from non-U.S. sources, it was felt that there would be little or no basis for embargoing the sale of the technology to the Soviet Union for military reasons.

The Second Theme: Control of Nonmilitarily Sensitive Technology

Several workshop participants were concerned that U.S. governmental preoccupation with the military aspects of the transfer of technology to the Soviet Union tends to obscure more fundamental, nonmilitary problems. Earlier in this chapter it was noted that the technology most sought by the Soviet Union is not so-called high technology, which is likely to be at the leading edge of military application, but rather proven factor savings technology. With certain exceptions (i.e., advanced computer-controlled machine tools, for example), this type of technology generally is not directly associated with advanced military applications.

As was developed in the fourth section of this chapter, there was concern among some workshop participants that the Soviet Union is able to obtain this technology from the West on terms that are favorable to the Soviets. In simplest terms, it was felt that the prices the Soviets pay for the technology can be considerably below the social value of the technology to the Soviets. No workshop participant believed that the United States should attempt to embargo or even restrict the sale of this technology to the Soviet Union. Rather, the concern was that a sufficiently high price be received. Some participants (but not all) were further concerned that the benefits to the United States of such sales be equitably distributed.

One dimension of the issue of the price of technology that requires underscoring, it was felt, is the multilateral dimensions of the issue. The technologies needed by the Soviet Union generally are available from a number of nations, mostly the larger OECD nations. It would thus be counterproductive for the U.S. government (or any other government) unilaterally to attempt to regulate the sale of this technology to the Soviet Union. The Soviets would conduct their business with nations not having regulations.

It was proposed that the nations that are major sellers of technology to the Soviet Union attempt to coordinate their policies regarding such sales. The process by which this might be accomplished, it was suggested, could be very similar to the one in which general tariff reductions were accomplished during the 1950's and 1960's. Representatives of the nations might sit down at a series of meetings to formulate a series of general principles to guide East-West trade. Principles might be established to guide policy with respect to sales of technology to the COMECON nations; extension of credit to these nations; pricing of large-scale importation of goods from the Soviet Union; and other problems associated with East-West trade. With respect to sales of technology to the Soviet Union, for example, it might be agreed that all sellers of a technology will price it to cover the full (social) costs of its development as well as the marginal costs. Once the principles had been formulated, the nations would agree to abide by them voluntarily.

An end result of this process might be the formation of an international agency that would have three major functions. The first of these would be to develop further the guiding principles of East-West trade. To this end, a permanent committee of representatives from each nation would meet at regular intervals. The second function would be to meet with representatives from the COMECON nations to discuss mutual problems associated with trade and to explore means by which mutual benefits from trade could be increased.

The third function would be to receive and hear complaints by nations or individual firms of violations of the agreed upon principles of East-West trade. For example, if a U.S. firm believed that it was losing sales to the Soviet Union because some non-U.S. firm was engaging in practices violating agreed upon principles, the firm could submit a grievance to the international agency (or have a grievance submitted through U.S. governmental channels). The agency could then determine if the grievance were meritorious and, if so, make recommendations for redressment. Any action on the recommendations, however, would be taken by individual governments; the agency, under this proposal, would not have the power to enforce the principles.

Under this proposal, neither the agency nor the represented governments would actually conduct trade with the Soviet Union. The actual conduct of trade would be carried out by the same institutions as are presently doing so. Thus, for example, private firms would continue actually to set the terms of U.S. trade with the Soviet Union. They would do so, however, in accordance with agreed upon principles, principles that would be formulated to ensure that the benefits of East-West trade did not flow unevenly to the East.

Discussion of this proposal was limited in extent and not all participants believed that the proposal was implementable. The consensus, such as it existed, was that the proposal should be further studied and explored.

Apart from exploration of the foregoing proposal, some participants stressed a need for more candid and rigorous attention to the underlying problems that the proposal is designed to meet. In the opinion of these participants, there has been a tendency among OECD governments, business enterprises, and scholars to pass too lightly over basic discrepancies in the structure of trade and investment between the Soviet Union and the OECD nations. The discrepancies should receive sustained and realistic attention. It would be ironic if the monopolistic and monopsonistic behavior of the Soviet Union's trade enterprises should draw the Western nations into an organization of comparable state enterprises on their own account, with a corresponding abandonment of the pluralistic market economy. These participants had no ready solutions of their own to offer, but they insisted on the importance of an open-eyed and resolute attention to the problem.

NOTES

1. For such a point of view, see M. I. Goldman, *Detente and Dollars: Doing Business with the Soviet Union* (New York: Basic Books, 1975) and Raymond Vernon and M. I. Goldman, "U.S. Policies in the Sale of Technology to the U.S.S.R." (unpublished, 1974). An analysis of Soviet price policy in Soviet-Eastern European trade is found in Martin J. Kohn, "Developments in Soviet-Eastern European Terms of Trade, 1971-75," in *Soviet Economy in a New Perspective* (Joint Economic Committee, 94th Congress, 2d Session, 1976), in which it was found that price changes in intra-CEMA trade instituted early in 1975 altered the terms of trade among CEMA's members (six Eastern European nations plus the Soviet Union) to the advantage of the Soviet Union. See also Franklyn D. Holzman, *Foreign Trade Under Central Planning* (Cambridge, Mass.: Harvard University Press, 1974) for a theoretical analysis.
2. See M. I. Goldman, "Autarchy or Integration: The U.S.S.R. and The World Economy," in *Soviet Economy in a New Perspective*, pp. 81-96.
3. See John Farrell and Paul Ericson, "Soviet Trade and Payments with the West," in *Soviet Economy in a New Perspective*, pp. 727-738.
4. See M. I. Goldman, "Autarchy or Integration: The U.S.S.R. and the World Economy." Also see Paul Ericson, "Soviet Efforts to Increase Exports of Manufactured Products to the West," in *Soviet Economy in a New Perspective,* pp. 709-726, for an account of Soviet efforts to increase exportation. Ericson reports that Soviet manufacturers often sell on world markets at major discounts. It is argued by one scholar that unexplored opportunities for procurement of commodities and manufactured products abound in the U.S.S.R. See Jack Brougher, "U.S.S.R. Foreign Trade: A Greater Role for Trade with the West," in *Soviet Economy in a New Perspective*, pp. 677-694.
5. The costs and benefits to the United States of trade with the Soviet Union are discussed in Henry Nau, *Technology Transfer and U.S. Foreign Policy* (New York: Praeger Publishers, 1976).
6. For views of this sort, see William Diebold, Jr., "The Role of the Western Multinationals in East-West Cooperation," in *Perspektiven und Probleme Wirtschaftlicher Zusammenarbeit Zwischen Ost-und Westeuropa* (Deutsches Institut Für Wirtschaftsforschung, Sonderheft 114, 1976), pp. 134-135, and John P. Hardt, "Summary," in *Soviet Economy in a New Perspective*, pp. ix-xxxix.

7. It has been noted that the next several years are critical to the Soviet Union as efforts are made to build up livestock herds and to replenish grain stocks. Shortfalls in grain production in 1976, 1977, or 1978 could lead to major problems, but bumper crops could propell Soviet agriculture to a hitherto unrealized growth path. See D. N. Carey, "Soviet Agriculture: Recent Performance and Future Plans," in *Soviet Economy in a New Perspective*, pp. 575-599.

8. See Joseph S. Berliner, "Prospects for Technological Progress," in *Soviet Economy in a New Perspective*, pp. 431-446, and Stanley H. Cohn, "Deficiencies in Soviet Investment Policies and the Technological Imperative," ibid., pp. 447-459. It might be noted that concurrent with deficiencies in process technology (associated with low factor productivity), the Russians have major problems in the domain of civilian product technology. The objectives of the Soviet Union include both improvements in factor productivity and an upgrading of the products produced.

9. Efforts to measure quantitatively the impact of technology transfer from the West upon Soviet economic performance have yielded mixed conclusions. The impact on the economy as a whole does not appear to be highly significant, although in certain sectors the impact may be quite large. A problem facing Soviet planners is that once a Western technology is acquired by the Soviet Union, the diffusion of the technology through the economy moves slowly if at all. For a survey of quantitative studies plus conclusions drawn from these surveys, see Philip Hanson, "International Technology Transfer From the West to the Soviet Union," in *Soviet Economy in a New Perspective*, pp. 786-812.

10. The long-run rate of growth of the Soviet economy does, of course, place some limit on the ability of the Soviets to expand military expenditures. To the extent that curtailment of sales of technology to the Soviet Union serves to reduce the long-term growth rate, the long-term capability of the Soviets to expand their military capabilities might also be curtailed. The workshop, however, felt that this argument was not a very powerful one for curtailment of commercial transactions with the Soviet Union.

For efforts to estimate quantitatively the long-term economic constraints on Soviet military expenditures, see Lars Calmfors and Jan Rylander, "Economic Restrictions on Soviet Defense Expenditure," in *Soviet Economy in a New Perspective*, pp. 377-393, and Hans Bergendorff and Per Strangent, "Projections of Soviet Economic Growth and Defense Spending," ibid., pp. 394-430.

11. Three agencies hold a major voice in matters pertaining to trade with the Soviet Union, and other agencies can become involved. The three agencies are the Department of Commerce, the Department of Defense, and the Department of State. The Executive Office of the President also is directly involved in such matters by way of the National Security Council.

Under U.S. law, the federal government can control export of goods if such control is indicated for reasons pertaining to national security, or if the good is in short supply, or if there are overriding foreign policy reasons to do so. In practice, control of trade with the Soviet Union comes through two channels.

The first and perhaps primary channel is the Bureau of East-West Trade of the Department of Commerce. This bureau must grant an export license to any firm domiciled in the United States that wishes to export to the Soviet Union or to any of the other so-called Communist bloc nations. Such a license may be routinely granted if the good to be exported is not on any proscribed or restricted list, notably the Commodity Control List or the Munitions Control List. If the Bureau of East-West Trade judges, however, that a good to be exported, although not on such a list, might in any way be vital to national security, it generally consults with the Office of International Security Affairs of the Department of Defense before granting the license. Even if Defense Department consultation is not actively sought by the Bureau of East-West Trade, the Secretary of Defense is empowered under the Jackson Amendment to the Export Administration Act of 1974 to review any export license application at his own initiative, and he is required by the Jackson Amendment to review any application involving items of direct military significance. Should the Bureau of East-West Trade decide to grant an export license over Department of Defense objections, the Department of Defense has the power (under the Jackson Amendment) to veto the license. The President can override the Department of Defense veto, but if the President should do so, Congress must be informed of the veto, the override, and the reasons for each.

If the good to be exported is on either of two lists of restricted goods, the COCOM List or the Munitions Control List, the Bureau of East-West Trade cannot issue an export license until the firm receives a license from the Office of Munitions Control of the Department of State (if the good is on the latter list) or a COCOM exception (if the good is on the former list).

The Office of Munitions Control is the second channel of control. A firm wishing to export any item on the Munitions Control List, a list of ordnance and other military-related goods, which is prepared by the Office of Munitions Control in consultation with the Departments of Defense and Commerce, must receive a license to export from the Office of Munitions Control *irrespective of the nation for which the export is intended*. Thus, the Office of Munitions Control is not exclusively concerned with matters pertaining to East-West trade, nor is the Munitions Control List really an "embargo list." It is claimed by some firms that have worked through the Office of Munitions Control, however, that in the domain of East-West trade the Munitions Control List is, for all practical purposes, a list of items proscribed from such trade.

The COCOM List is a list of items that the COCOM nations (which are the nations of NATO, less Iceland plus Japan) jointly embargo from export to the Communist nations. In principle at least, all COCOM nations voluntarily adhere to the same embargo list. Neither the United States nor any other COCOM partner, however, is bound either by law or by treaty to enforce the embargo. But the fact is that the embargo is almost always adhered to by all COCOM nations, and instances of a COCOM nation unilaterally violating the COCOM agreements have been rare.

There are provisions for obtaining exceptions to the COCOM embargo. Should the United States, for example, seek an exception (generally, the decision to seek an exception would have to be approved by the Department of State, the Department of Commerce, the Department of Defense, and the National Security Council), the Department of State takes the matter up before the COCOM permanent delegates in Paris, who in turn send the Department of State officials to the export control agencies in their respective governments. The exception must be granted unanimously by the COCOM nations.

Should COCOM decide not to approve an exception, the nation seeking an exception may overrule COCOM by declaring a national interest exception. Only the head of state (for the United States, the President) can declare a national interest exception.

12. See *An Analysis of Export Control of U.S. Technology-- A DOD Perspective: A Report of the Defense Science Board Task Force on Export of Technology* (Washington, D.C.: Office of the Director of Defense Research and Engineering, U.S. Department of Defense, February 1976).
13. It was recognized by the formulators of these

proposals that they would have little meaning if technologies being controlled by the United States were freely available to the Soviet Union from other Western nations. This problem, it was suggested, could be countered in two ways. First, the preferred way, the large Western industrialized nations could multilaterally adopt and implement the proposals and thus jointly control export of technology to the Soviet Union. Second, the United States could move to block the transfer of keystone technologies to all nations, or at least to those showing a willingness to transfer these to the Soviet Union.

It was believed by the majority of the workshop that the first of these two approaches was not feasible and that the second would be ineffective.

6 TECHNOLOGY AND TRADE ISSUES RELATING TO DEVELOPING NATIONS

Confronting the United States today are a series of demands by the developing nations through the Group of 77 for changes in the world economic order. A major component of these demands is a call for an acceleration of the rate of technology transfer to developing nations and for changes in the terms upon which developing nations receive technology. Much of the workshop discussions were concerned with how the United States should address these demands.

It was recognized that the underlying cause of the demands of the Group of 77 is the almost universal objective of developing nations to increase their share of world income. Most of these nations see industrialization as one necessary means to fulfill this objective, and a stated goal of the Group of 77 is for the developing nations to increase their share of world manufacturing output from about 8 percent at the present time to 25 percent by the year 2000.

1. PROSPECTS FOR INDUSTRIALIZATION OF THE DEVELOPING NATIONS

The fact appears to be that of the more than 100 nations that can be categorized as "developing," only a relative handful seem to be on the road to industrialization. Several of these, such as Brazil, Iran, India, Mexico, Taiwan, and South Korea, are already quite industrialized, and it might no longer be strictly accurate to consider these as developing nations. The extent and rate of industrialization of most developing nations, however, lags far behind that of the few high performers.

Exactly why a nation should not be able to achieve industrialization is a complex issue that is not fully understood. A large number of inter-related factors appear to determine whether or not a nation is industrialized. Some of these factors are basically economic in nature, but equally important factors are sociological. It has been noted by many development economists that internal market size is an important economic factor in the development of a modern industrial sector, and it follows that those developing nations with internal markets large enough to support industries that can produce at minimum efficient scale are more likely to achieve industrialization than those without such markets. In fact, of those developing nations that are on the road to industrialization or at least have the potential for industrialization, most are characterized by large internal markets or potentially large internal markets. A smaller nation, one whose market cannot internally absorb the output of efficiently scaled industry, must rely on exportation if it is to develop an internationally competitive industrial sector. A few small developing nations, Taiwan and South Korea for example, are apparently succeeding in developing manufacturing industries that are centered on exportation. The prospects for industrialization of many developing nations, however, are hampered by the small size of internal markets.

An important factor affecting the ability of a nation to industrialize is its infrastructure, including both its physical infrastructure (i.e., transportation and distribution systems, electrical power systems, communications systems, etc.) and its social infrastructure (i.e., skill and educational attainments of the population, nature and quality of social organizations, etc.). It was noted by the workshop that a major problem common to many developing nations is a lack of that cadre of trained engineers, designers, technicians, and craftsmen required by a nation if it is to be industrialized. It was also noted that the development of such a cadre on a large scale would be difficult to achieve in a nation were there not a growing industrial sector to provide employment opportunities for persons newly entering the ranks of the skilled.

The total list of factors believed to affect the rates and levels of industrialization of developing nations is a long one and will not be elaborated upon at any length here.[1] In addition to the problem of small internal market size, developing nations face numerous obstacles to industrialization, including low per capita income, low average levels of education among the population, low rates of capital formation, low factor productivity in the agricultural sector, lack of physical and social infrastructure, and (in some nations) problems of widespread malnutrition and poor health. The relative importance of these obstacles varies widely from nation to nation, and about the only general comment that can be made is that for a large number of nations there has been some slow but measurable progress in overcoming these obstacles over the past two decades or so.[2]

Of the many obstacles to industrialization faced by developing nations, access to technology per se has not until quite recently been articulated as a major problem. The types of technology required most urgently by the developing nations are generally mature technologies that are quite widely in use in the industrialized nations. Three problems associated with access to technology, widely articulated by developing nation spokesmen, are as follows: (1) technology is transferred to developing nations by multinational corporations on terms that are unfavorable to the host nations; (2) the international patent system acts to impede acquisition of industrial technology by developing nations; and (3) the technology is not adapted to reflect the differences in relative factor costs between the donating and host nations. Little empirical work has been presented to test the validity of these assertions, although efforts to do so are being mounted.[3] The assertions do, however, form part of the basis for the current demands of the Group of 77.

Despite the many problems associated with industrialization of developing nations, it should be noted that the rate of increase of industrial output of some of these nations has outpaced that of most of the large industrialized nations in recent years. (See Table 12.) Concurrently, export of manufactured goods by developing nations has been increasing at rates exceeding the overall expansion of their economies. Much of this exportation has been to industrialized nations. (See Tables 13 and 14.) A disproportionate amount of these exports, however, originated from only a few nations, such as Mexico, Brazil, Taiwan, South Korea, India, and Iran.

Despite the growth of manufacturing industry in developing nations, the rate of gross investment of developing nations as a whole has been less than that of industrialized nations. (See Table 15.) The net consequence of this, it was felt, is that in the long run, if the net investment rate differential persists, it will be difficult for developing nations to grow as fast as industrialized nations.

TABLE 12 Index Numbers for 1973 of Manufacturing Output for Selected Nations (1970 = 100)

Nation	Index Number	Nation	Index Number
United States	119	Panama	126
West Germany	113	Argentina	123
Japan	130	Colombia	129
United Kingdom	110	Peru	128
Egypt[a]	115	Venezuela[a]	112
Ghana[a]	129	India	108
Kenya[a]	120	Iran[a]	137
Morocco	121	Republic of Korea	187
Tunisia	125	Pakistan	109
Costa Rica[a]	126	Philippines	134
Guatemala	122	Syria	128
Mexico	122		

[a] Data from 1972.

SOURCE: United Nations, *Statistical Yearbook, 1974*.

TABLE 13 Exports and Imports of Developing Nations, by Major Commodity Class and by Destination, 1973 (U.S. $ billions)

Exports of Developing Nations	Importing Region		
	All Areas[a]	Developed Nations	Developing Nations
Food products	17.94	14.49	3.45
Raw materials[b]	16.95	13.69	3.27
Fuels	41.97	34.65	7.32
Chemical products	2.00	0.91	1.09
Machinery[c]	4.81	3.32	1.49
Other manufactures	17.96	13.79	4.16
Subtotal: manufactured goods[d]	24.77	18.02	6.74
TOTAL	102.24	81.29	20.94

[a] Except Communist nations.
[b] Excluding fuels.
[c] Includes transportation machinery.
[d] Excludes processed foods.

SOURCE: United Nations, *Yearbook of International Trade Statistics, 1974*.

TABLE 14 Index Numbers of Exports of Manufactured Goods by Developing Countries in Trade with Developed Countries (1963 = 100)[a]

Year	Index Number	Year	Index Number
1960	81	1967	137
1961	82	1968	166
1962	90	1969	195
1963	100	1970	211
1964	112	1971	220
1965	115	1972	291
1966	130	1973	334

[a]Excludes processed foods.

SOURCE: United Nations, *Yearbook of International Trade Statistics, 1974*.

TABLE 15 Gross Investment of Developing Nations and Industrialized Nations as a Percentage of GNP[a]

	1966-1970	1971	1972	1973
Developing nations	19.4	20.5	20.8	21.1
Industrialized nations	22.1	22.4	22.4	23.3

[a]Developing nations having loans from the World Bank only.

SOURCE: International Bank for Reconstruction and Development, *Annual Report*, 1975.

2. TECHNOLOGY AS A CONSTRAINT UPON THE INDUSTRIALIZATION OF DEVELOPING NATIONS

In international forums, such as the United Nations General Assembly and the UNCTAD IV meetings, the developing nations have presented a common front with respect to the articulation of their problems regarding the transfer of industrial technology to their economies. The allegations are (1) the cost of technology, especially as manifested in the form of direct foreign investment by multinational corporations, is too high; (2) the international patent system impedes excessively efforts by developing nations to acquire industrial technology; (3) the technology is not appropriate to the relative factor endowments of the importing nations, the technology being too capital intensive; and (4) in conducting business in developing nations, multinational corporations engage in "unfair" practices, including the creation of artificial barriers to entry to local entrepreneurs in industries in which, were "fair" conditions to prevail, the local entrepreneurs could compete successfully. These allegations form the basis for developing nations' demands that there be a mandatory "code of conduct" for the transfer of technology by multinational corporations.

Two comments were noted by the workshop with respect to these allegations:

First, it was felt that the nations that are most vociferously articulating the complaints are a few of the most industrially advanced of the developing nations. As Table 16 suggests, these nations account for the majority of U.S. private direct investment outside of the industrialized world. The sentiment was expressed that these nations are not advancing the concept of a code of conduct with the expectation that their demands will be fully met but rather with the hope that by doing so they can improve their bargaining position in relation to multinational corporations.

Second, it was expressed that efforts to develop products and technologies that are more appropriate to developing country conditions have in the past had little success. The primary problem with such development has been the cost of doing so. It was expressed that developing countries' perceptions of the cost of adapting process technology to changes in relative factor input costs are unrealistically low.[4]

The bulk of the workshop discussion of the demands of the developing nations with respect to transfer of technology centered on the process by which the United States should deal with the demands, rather than with the merits or demerits of the demands themselves. In this regard, four options were discussed:

1. To what extent, if any, should the United States attempt to coordinate its policies with respect to the transfer of technology to developing nations with the corresponding policies of other OECD nations? To what extent should there be consultation among the OECD nations over such policies? Is the present approach of the OECD, to develop its own code of conduct as an alternative to the Group of 77 code, a fruitful one?

It was noted that the developing nations have displayed an unparalleled degree of unity in presenting in international

TABLE 16 Private Direct Investment Position of the United States in Developing Nations at Year-End, 1975 ($ millions)

Nation	Mining and Smelting	Petroleum	Manufacturing	Other	Total
Brazil	131	292	3,105	1,035	4,563
Mexico	80	21	2,433	643	3,177
Bermuda	--	110	--	2,839[a]	2,949
Venezuela	NA	861	678	NA	2,065
Panama	-1	105	122	1,599	1,825
Bahamas	NA	NA	96	680[a]	776
Colombia	17	62	380	189	648
Jamaica	NA	NA	219	NA	655
Other Western Hemisphere	NA	NA	1,520	NA	5,565
Africa	486	1,337	231	343	2,397
Middle East	5	3,673	164	666	4,508
Indonesia	NA	1,298	94	NA	1,612
Philippines	NA	135	339	NA	733
India	--	79	254	31	364
Other Asia	NA	1,254	802	NA	3,037
TOTAL	2,144	11,337	10,437		34,874
Western Hemisphere	1,472	3,370	8,553		22,223
Africa	486	1,337	231		2,397
Middle East and Asia	186	6,630	1,653		10,254

NA = not applicable.
[a]Estimated.

SOURCE: U.S. Department of Commerce, *Survey of Current Business*, August 1976.

forums a set of demands to the industrialized nations with respect to transfer of technology from the industrialized to the developing nations. The issue then boils down to the following: given that the OECD nations face a united bloc when dealing with the developing nations, should the OECD nations organize themselves into a counter bloc?

A general consensus in the workshop was that the answer to this last question, for any number of reasons, is no. The developing nations, it was noted, are a widely heterogeneous group of nations having much different interests and priorities. The major unifying factors among them have been feelings of powerlessness in the world as it exists today and of antagonism toward the rich nations. For the OECD nations to confront the developing nations on a bloc basis would be to exacerbate the antagonisms and to deny the plurality of the developing nations.

The sentiment of the workshop was that discussions between the Group of 77 and the industrialized nations should be reduced from the present level of quasi-confrontation to the level of a useful dialogue. The opinion of most workshop members was that the demands of the Group of 77 have been advanced as an opening "bid" in a process seen by Group of 77 leaders as essentially a bargaining game. Real progress, it was felt, could be made only through a dialogue that explores possible areas of conciliation and accommodation.

The workshop did believe that the U.S. government, in addressing Group of 77 demands, should consult with other OECD nations (1) to determine if there is any common agreement among OECD nations over which particular demands of the Group of 77 might be acceptable to the industrialized nations as a whole; (2) to determine if there is any common agreement over what counterproposals the industrialized nations should make to the Group of 77; and (3) simply to learn and remain abreast of the thoughts and positions of other OECD nations. This consultation, it was believed, should be largely informal but fairly extensive, involving business and labor leaders of the OECD nations as well as governmental representatives. It was noted that a certain amount of this sort of consultation was already going on among OECD nations.

2. Can or should the U.S. government collect and make available to developing nations comparative information regarding U.S. suppliers of technology?

It was noted by the workshop that some of the industrial

technology needed by developing nations is readily available from many sources in the industrialized nations and that its availability is not subject to patents or other restrictions on its application. The belief was that developing nation institutions do not know how to locate and evaluate alternative suppliers of these technologies. This led to the suggestion that the U.S. government might therefore collect comparative evaluatory information on U.S. private sector suppliers of technology and disseminate this information to developing nations. Several participants believed that this would enable developing nations to obtain access to technology on more favorable terms than at present. It was felt by some workshop participants that such a role by the U.S. government would serve as one step toward accommodating the demands. It was pointed out that to some extent the governments of other OECD nations do provide this service, but that no equivalent service is performed by any U.S. agency.

A number of participants, however, registered disagreement with this proposal. It was felt that the amount of information that would have to be generated to enable the U.S. government to fulfill this role usefully would be absolutely immense. Thus, it was not clear to the workshop that the task of accumulating comprehensive comparative information was feasible.

Those who disagreed with the proposal felt that such a role for the U.S. government would necessarily put the government in a position of having to recommend specific suppliers to specific nations. It was believed that this would amount to undue governmental influence in the competitive process and could raise potential questions about conflict of interest between the government and private entities. It was noted that this conflict-of-interest issue could be avoided were the government simply to catalog all suppliers of a technology without comparatively assessing them. However, such a listing, it was felt, would be of little use to the developing nations.

There were differences of opinion as to whether or not the government should even take limited action in supplying comparative information. Some believed that the government should remain out of this area altogether and that the marketplace should be the sole determinant of what technology is located where. Others believed that the government might assist developing nations on a case-by-case basis to locate and evaluate different suppliers of technology. It was pointed out that to a very limited extent, the U.S. Agency for International Development already provides services of this sort.

Those workshop participants who favored the proposal made the following points: (1) the developing nations want and need the best technology on the best possible terms; (2) many of the other OECD nations (Japan and Germany were specifically mentioned) have institutional mechanisms designed to meet these needs; and (3) the free market approach, as officially espoused by the U.S. government and supported by most workshop participants is no longer viable. As evidence of this last point, these conferees noted that even private deals cannot be consummated in developing nations without governmental approval at one level or another.

3. Should the U.S. government provide additional incentives for U.S.-based firms to invest in developing countries?

There was no discernible consensus in the workshop on this issue. On the one hand, it was felt by some participants that additional incentives would make it possible for private firms to operate in developing nations on terms that would be more favorable for the host nations' economies. On the other hand, it was expressed by representatives of organized labor that existing incentives for U.S. corporations to transfer operations to developing nations are already too strong.

One position, taken by several workshop participants, was that the best policy for the U.S. government in this regard was strict neutrality. The reasoning in support of this viewpoint was that the U.S. government currently is caught between developing nations' demands on the one hand and demands of U.S. labor on the other. The belief was that these two sets of demands call for contradictory policies, and that, therefore, the best overall position for the U.S. government is a neutral one.

It was pointed out, however, that strict neutrality would imply the following:

1. No incentive or disincentive in the tax system for investment in the developing nations by U.S.-based corporations.
2. No investment guarantees of any sort.
3. No preferential tariff treatments for imports originating from developing nations.

There was a significant dissent from this position. Most participants believed that the U.S. government should continue the system of generalized tariff preferences now in place. Several advocates of neutrality also advocated an active U.S. policy to transfer labor-intensive technologies to developing nations. There was little agreement, however, on how such an active policy could be pursued without creation of special tax incentives or investment guarantees, moves that would directly violate the concept of neutrality.

Most workshop participants believed that there should be some form of official protection for U.S. companies against expropriation of their assets by host nations' governments. Advocates in the workshop of U.S. governmental neutrality felt that this protection should take the form of agreements between developing nation governments and private foreign investors over the terms upon which the foreign investors could operate in developing countries without being expropriated. It was felt that phase-out agreements, whereby foreign investors agree to relinquish ownership of assets in developing nations over time, might constitute one basis for establishing such terms. It was pointed out, however, that under a phase-out agreement, the foreign investor has an incentive to allow assets to depreciate fully in the phase-out years of the contract, and this in turn generates pressure on the part of the host nation government to expropriate the assets prematurely.

Other participants felt that if private direct investment is to continue to flow to developing nations from the United States, there must be U.S. government incentives to promote it. However, they did not suggest that the incentives should be incorporated into the tax system, even if present tax incentives are minimal. The case was made that because of the investment tax credit for domestic investment, the tax system actually provides a disincentive for foreign investment. Rather, it was felt that incentives should take the form of U.S. government insurance against expropriation risk, it being stated that the present Overseas Private Investment Corporation system is inadequate.

The discussion on this topic reflected an issue that will necessarily intervene in any effort by the United States to formulate policy to deal with the Group of 77 demands: the possibility that any effort of the U.S. government to meet these demands will meet with opposition from U.S. labor. It was noted that the developing nations, including the United States, seek to become more competitive in international markets for manufactured goods. To the extent that this is seen by U.S. labor as inimical to labor's interest, labor is likely to stand in opposition to U.S. policy to promote the interests of developing nations.

4. Should there be elimination of enforcement of patent protection for technologies sought by developing nations?

One stated objective of the Group of 77 is to dissolve the international patent system within the developing world and to require holders of patents to make their knowledge available to developing nations free of charge. This objective is based on the allegation that the patent system impedes or excludes developing nations from acquiring needed technologies and creates a dependence by these nations upon obtaining needed technologies from patent holders in the industrialized world.

Although some sympathy was expressed for the developing nations' allegations on this issue, there was no enthusiasm within the workshop for major modification of the patent system. It was generally held that the protection provided to technological innovators by the patent system is necessary to the generation of new technology. It was also held that dissolution of the patent system within the developing world could not be accomplished without a dissolution within the industrialized world as well.[5]

3. RECOMMENDATIONS

The workshop recommended several specific proposals with respect to how the U.S. government might respond to Group of 77 demands. It must be noted, however, that not all of these proposals were unanimously approved by all participants. In particular, of the five proposals presented in this section, the last two were endorsed by only a minority of the workshop participants and are not presented as representing majority views.

1. That an analysis be undertaken of the policy implications to the United States of demands by developing nations with respect to the transfer of technology.

It has already been noted that the workshop did not discuss the specifics of the demands of the Group of 77. It was felt rather that a major study should be undertaken to analyze and report on the implications of these demands. It was felt that one important aspect of this study should be to assess the impact on the U.S. economy of industrialization of developing countries. Of particular concern would be the evaluation of codes of conduct proposed by the Group of 77 for international transfer of technology, including the identification of the aspects of proposed codes that would be unacceptable to the United States, the aspects that would pose major difficulties, and the areas that would allow for mutual accommodation between the developing nations and the United States. In making this sort of assessment, account should be taken of the views of other OECD nations. It was noted, for example, that the OECD secretariat has established committees to study the Group of 77 draft code of conduct and to formulate an alternative OECD code of conduct. It was felt that the proposed standing committee should work closely with the OECD secretariat committees.

The undertaking of such a study was endorsed by a large majority of the workshop participants. It was pointed out to the conferees by a representative of the U.S. Department of State that the U.S. official position regarding codes of conduct has undergone a certain amount of change during 1976. In January the U.S. government was officially unwilling even to discuss the possibility of official support for a code of conduct of any sort--voluntary or mandatory; by summertime, however, there was talk of the possibility of some sort of a U.S. official support of a voluntary code. The representative suggested that it might be most useful for the U.S. government to be able to turn to an advisory body for consultation on the future course of U.S. policy in this direction.

2. That there be created mechanisms by which nonproprietary technologies could be transferred more readily to developing countries.

The argument for this proposal was that there is much technology that bears directly on many of the problems of developing nations and that is not proprietary. In some cases it is in the public domain because it was largely generated by the government. In other cases it is in the open technical literature and is not covered by patents. Examples of the former can be found in the areas of agriculture, food processing and distribution, water management, energy, housing construction, highway construction and maintenance, civil aviation, and occupational health and safety. Examples of the latter can be found in every area of industrial technology. It was felt that this vast body of technology could be utilized to aid developing countries and that there is insufficient recognition of the amount that is available.

Several questions were raised regarding this proposal, most notably the following: (1) Is not the task of identifying this technology already being accomplished by the U.S. Agency for International Development (AID) or other U.S. governmental agencies? (2) If the task is not already being accomplished, what specific additional mechanisms must be established?

A response to the first question was that although U.S. AID does fund projects that involve U.S. government personnel from other governmental agencies traveling to developing nations and making their services available to the projects, these are of a rather limited nature. In the opinion of U.S. governmental representatives speaking before the workshop, the store of technology residing in federal departments and agencies is not generally readily available to developing nations through U.S. AID projects. Several of these representatives felt that greater accessibility might be achievable only through legislative mandate to allocate responsibility to one agency (possibly, but not necessarily, AID) for coordinating an interagency program for transfer of government-controlled technology to developing nations. Most participants endorsed the proposition that any legislative mandate that might be required to give a coordinating agency the ability to carry out such a program should be enacted.

With respect to nonproprietary technology that does not derive from public bodies, the problem is rather different. In this case it is primarily a matter of identification of sources of information and collection of information. This might be very difficult for a developing country to accomplish and yet be quite manageable for knowledgeable U.S. technologists.

It was recommended by the workshop that the United States should work with other OECD nations to coordinate programs for transfer of nonproprietary technology to developing nations. It was felt that there should be some sort of broad division of labor among the OECD nations to carry out this task. The OECD itself could act as an information and guidance center for the developing nations, assisting these nations in evaluating available technology.

It was noted that transfer of nonproprietary technology would mostly be in the interests of the poorer developing nations, those that are not experiencing significant progress toward industrialization. The more rapidly industrializing nations already are self-sufficient or nearly so in these technologies. Because some of these nations might be developing technologies that are especially suitable for developing nation conditions, it was felt that the assistance of nations such as Brazil might be sought in the effort to transfer nonproprietary technologies to poorer nations.

3. That the United States officially support and financially assist regional institutions, to be located in developing nations, whose goal would be the location and development and application of technology appropriate to the conditions prevailing in the developing countries.

This proposal, which was recommended with some reservations by the workshop, calls in effect for the creation and development of centers of applied research in developing nations. It should be noted that this proposal directly addresses one frequent demand of the developing nations, expressed through the Group of 77, notably that R&D capability be transferred to the developing world.

The reservations of the workshop regarding this proposal are worth noting. It was observed that a number of developing nations have attempted to create governmentally sponsored research laboratories within their borders. The experience in most of these cases has been that the efforts of the laboratories generally become focused upon pure scientific research. This sort of research in developing nations has produced impressive results in some instances when measured in terms of articles published in prestigious research journals or honors in the scientific communities. It has generally not been the case, however, that these laboratories have produced technological innovations utilizable by the local economy. There are two reasons for this. First, the development of applicable technology requires entrepreneurial skills, skills that are not usually attracted to governmental research centers in any national environment. Second, the people who typically do make up the professional staffs of research institutions of this type are most often ones whose values are much more oriented toward the basic problems confronted in pure research than toward the application-orientation of the applied research laboratory.

Thus, although the workshop did recommend official U.S. support of regional R&D institutions, it recognized that a major effort might be required to ensure that these institutions work on problems relevant to the developing nations. Such an effort would be fruitless unless the institutions were somehow directly linked to the productive sectors of developing nations' economies. One possible approach discussed was the establishment of a major international institution to carry on applied research on the processes of industrialization. Such an institute would serve as a center for the study of the many facets of industrialization in the developing world. It was felt that no institution presently exists to serve this role.[6]

In addition to financial assistance, the workshop believed that the U.S. government could give these institutions assistance in the establishment of linkages to U.S. R&D institutions.

4. That the U.S. government render technical assistance to developing nations to develop internal safeguards with respect to the introduction of technology.

This proposal was made largely in response to the Group of 77 allegation that multinational corporations often engage in practices in developing nations that would be illegal in the parent company nations. The demand of the Group of 77 is that the governments of the industrialized nations in which the multinational corporations are based take action to prevent the companies from engaging in these practices. The majority of the workshop believed that for the U.S. government to attempt to regulate U.S. corporations in this manner would be impractical and would raise serious questions about the extraterritorial powers of the U.S. government over the conduct of its citizens, corporate or otherwise. In fact, for the federal government to attempt to regulate the conduct of U.S. corporations on foreign soil would almost certainly require its intrusion upon the sovereignty of other nation states, a practice that the U.S. government is already accused of too often. The consensus of the workshop was that it is the task of the host nation government to regulate the conduct of multinational corporations operating within its domain and not the task of the U.S. government. It was expressed by some workshop participants, however, that it would be quite proper for the U.S. government to assist developing nations to develop the necessary internal capability to regulate effectively foreign corporations operating on their soil.

The specific areas in which U.S. corporations allegedly have engaged in nondesirable practices in terms of technology transfer have been the following: excessive product proliferation (introduction of too many products designed to perform the same task); introduction of products that might be harmful to the consumer without adequately warning the consumer of the potential hazards (this has generally been associated with pharmaceutical products); use of production processes that are dangerous to production workers or to inhabitants of areas close to the plant site; and failure to install safety or pollution control equipment that would be mandated in the United States. The judgment of the workshop was that all of these actions could be controlled effectively by host nation governments.

It should be noted that the creation of a regulatory environment in the developing nations at least as stiff as that prevailing in the United States was an idea endorsed by U.S. labor participants in the workshop. A frequently voiced complaint of U.S. labor is that U.S. corporations can move production out of the United States and into developing nations in order to escape from the effects of U.S. regulation. If the nature of regulation were everywhere to be the same, of course, there would no longer be any incentive for shifts in production locations to be made in response to differing regulatory requirements.

5. That the U.S. government officially encourage "coproduction" agreements, whereby U.S.-based firms would manufacture in developing nations goods for export to the U.S. market.

This proposal was supported by many of the workshop participants, but a number disagreed. Not surprisingly, perhaps, strong dissent was registered by the representatives of U.S. organized labor.

The major argument for this proposal was that developing nations are demanding greater access to industrialized nations' markets and that without such access the developing nations will not be able to develop within their industrial sectors the scale economies necessary to make these sectors internationally competitive. The major counterargument was that through existing preferential tariff arrangements the developing nations already have favorable access to U.S. markets and that imports from developing nations already are doing enough harm to domestic U.S. industry.

To a very great extent, the disagreement over this particular proposal summarized the entire debate within the workshop over the matter of technology transfer to developing nations. The central theme of this debate was the balancing of the benefits to the United States, resulting from such transfer, against the costs. The benefits, as has been noted, are mostly long term in nature and are manifested in both political and economic forms: the political benefits of raising the income levels of the majority of the world's population that live largely without the benefits of modern technology and the economic benefits that might accrue from trade conducted between the United States and nations that came to realize their potential comparative advantage. The costs are shorter term in nature but fall upon particular segments of the U.S. population in a disproportionate way. Long term benefits to the United States mean relatively little to a worker whose job has been eliminated and whose skills are no longer needed by anyone. It was recognized by the workshop that a balancing of these costs and benefits is a delicate political process. It was also recognized by the workshop that the recommendations that it was prepared to endorse do no more than partially address the dilemma.

NOTES

1. Numerous treatises have been written on this topic. See, for example, E. E. Hager, *The Economics of Development* (Homewood, Ill.: Richard D. Irwin, 1968) for a survey of the problems facing the industrialization of developing nations.
2. See *Trends in Developing Countries* (Washington, D.C.: World Bank, 1973), chaps. 3-6.
3. See, for example, S. A. Morley and G. W. Smith, "Limited Search and the Technology Choices of Multinational Firms in Brazil," *Quarterly Journal of Economics*, May 1977.
4. For a discussion of appropriate technology, see Richard Eckaus, *Appropriate Technologies for Developing Countries* (Washington, D.C.: National Academy of Sciences/National Academy of Engineering, 1977); see also N. Jecquier, editor *Appropriate Technology: Problems and Promises* (Paris: OECD, 1976); H. Pack, "The Substitution of Labour for Capital in Kenyan Manufacturing," *Economic Journal*, March 1976; R. F. Solomon and D. J. C. Forsyth, "Substitution of Labour for Capital in the Foreign Sector: Some Further Evidence," *Economic Journal*, June 1977; Louis T. Wells, "Economic Man and Engineering Man: Choice of Technology in a Low Wage Country," *Public Policy*, Spring 1973.
5. For a fuller discussion of these issues, see H. G. Johnson, "The Efficiency and Welfare Implications of the International Corporation," in C. P. Kindleberger, editor,

The International Corporation (Cambridge, Mass.: MIT Press, 1970).
6. See *Meeting the Challenge of Industrialization* (Washington, D.C.: National Academy of Sciences/National Academy of Engineering, 1973) for a detailed proposal for such an institution.

APPENDIX A COMMENTARIES ON THE CHANGE IN THE RELATIVE ECONOMIC STATUS OF THE UNITED STATES WITH RESPECT TO OTHER OECD NATIONS

That the United States is in some sense losing ground in the development of new technology to other advanced nations is antithetical to notions that were popular during the 1950's and 1960's. Up until perhaps 1967 or so, it was popularly assumed that there existed a large, virtually insurmountable "technology gap" between the United States and the rest of the Western industrialized nations. It was claimed that the sheer size and level of technological sophistication of the U.S. market was such that U.S. companies in some industrial fields might develop technological capabilities so great that European competitors would permanently lose their competitive positions in world markets.[1]

It was during the last years of the 1960's that this theme of an irreversible technology gap began to attenuate and a new theme of a loss of the U.S. lead in technology began to appear. A growing number of scientists, engineers, economists, and political scientists have commented upon this new theme. This section explores the ideas of a few of these.

One of the most senior of the commentators is Charles P. Kindleberger.[2] Kindleberger speculates that during the next 25 years or so, the United States might enter a phase of economic history not unlike that experienced by the United Kingdom during the late nineteenth century. In the eighteenth and nineteenth centuries, Great Britain led the world into the industrial revolution, and during much of of that time Britain was the acknowledged source of the majority of the world's industrial innovation. The latter half of the nineteenth century, however, saw the leadership position of Britain begin to erode as other, more dynamic nations (especially the United States and Germany) first began to equal and then to surpass Britain's industrial

accomplishments. Britain, as Kindleberger colorfully puts
it, entered a "climacteric"--literally, a "change of life."
Britain's response to the challenge of a declining position
was simultaneously to become defensive and to become a
rentier. Rather than attempt to rejuvenate the technological
capabilities of her economy, Britain simply tried to hold
her own in those areas in which she had historically excelled
(textiles, metallurgy, machinery, and shipbuilding) and did
not succeed at that. Rather than reinvesting at home, the
British invested large sums overseas and lived off the
earnings--thus, the capital base of Britain deteriorated
even as the overseas earnings grew.

Kindleberger sees similarities between the British position in the late nineteenth century and that of the United
States in the late twentieth. The United States is investing substantial sums overseas, as did the British. In many
industries in which the United States has clearly possessed
a world lead in industrial technology for a large portion
of the present century (machine tools, automobiles, petroleum refining, and metallurgy would be examples), the lead
seems to be eroding (or, at least other nations are catching up). In some of these industries, a defensive posture
by U.S. industry is evident: rather than redoubling efforts
to generate new innovations, firms in many of these industries instead have actually curtailed investment in innovation and have called for governmental protection against
imports. These tendencies are viewed by Kindleberger as
being particularly ominous for the long-term U.S. balance
of trade. The "comparative advantage" of the United States
historically has been to create new products through innovation and to export these products. (See the discussion
of technology and trade in Chapter 2.) These new products
have continuously displaced older products as the "leading
edge" of U.S. exports. If the flow of new products ceases,
in Kindleberger's point of view, the likely consequence
is to be a deterioration in the U.S. terms of trade.

One element in common to both Great Britain in the late
nineteenth century and the United States at the present
time is, in Kindleberger's analysis, an increasing propensity to consume. A possible consequence of this is a declining rate of investment, either in tangible productive
capital or in innovation. The society in effect "lives
off capital," a viable thing to do in the short run but
disastrous in the long run.[3]

Kindleberger acknowledges clear differences between
the present U.S. position and the past British position,
however. United States foreign investment has to some

very large extent been direct investment, wherein U.S. firms invest abroad to exploit technological (or other) advantages, while British investment overseas in the nineteenth century was largely passive portfolio investment.[4] In the very high technology industries such as telecommunications, aerospace, and electronic computation, the United States appears to have retained a significant lead over virtually all other nations; such a lead in the then high technology industries was not evidenced by Great Britain during the late nineteenth century, the lead in such industries having passed to Germany and the United States early on.[5] If, in fact, the United States is losing the lead, Kindleberger is unsure to whom the technological lead is passing. In the case of nineteenth century Great Britain, it was clear that by 1900 the lead had moved to the United States and Germany, both of which were large nations better endowed with natural resources than was Britain. Today, when one attempts to think of the nations that are most strongly challenging the United States in terms of industrial innovation, Japan and Western Germany come most immediately to mind. Japan, however, is a nation whose primary strength has been to imitate the industrial innovations of others, improving the design of products that are known to be commercially viable and to manufacture them more efficiently than anyone else. Grass roots innovation has not been a Japanese strength, and in the views of some analysts, if future Japanese economic growth depends upon original innovation, Japan might well experience difficulties.[6] Germany's strengths have lain in the areas of producing products, especially durable goods, of high quality, and in some industries (most notably chemicals), Germany has excelled in industrial innovation.[7] However, both Germany and Japan are relatively small nations with limited natural resources, and the economies of both nations are having difficulty simply finding the physical space in which to expand.[8] Thus, Kindleberger's views are somewhat ambiguous. On the one hand, he sees the possibility that the United States is entering a climacteric to be a very real one. On the other hand, he is not quite sure to whom, if anyone, the role that has been filled by the United States during the post-World War II era is likely to pass.

Sharing Kindleberger's concerns over a possible U.S. climacteric are numerous other scholars and practitioners. Robert Gilpin, for example, has written extensively on the impending demise of the U.S. economy, advancing the hypothesis that technology transfer abroad by U.S. multinational corporations is the culprit.[9] Gilpin's hypothesis is that multinational corporations, seeking faster-growing markets offering higher profit margins than those of the United States, increasingly invest outside the United States. In doing so, these firms use their latest and most advanced technologies in their overseas operations and thus transfer their best technologies to foreign markets. By doing so, Gilpin argues, U.S. multinational firms contribute to the growing dynamism of foreign economies, which ultimately come to outperform the United States economy. While in many ways similar to the conjectures of Kindleberger, Gilpin's hypotheses raise the possibility of technology transfer by U.S. multinational firms as being one major agent behind the deterioration of U.S. technological leadership, a possibility not emphasized by Kindleberger.

One of the protagonists of the view that the United States is losing the technological lead is Michael Boretsky.[10] Boretsky stands as perhaps one of the most pessimistic commentators on the state of U.S. technology. His published view is that the rate of U.S. innovative activity has declined perilously and that this decline is manifested in trends toward reduced U.S. factor productivity and "unfavorable" U.S. trade balances. Boretsky's views were discussed by the workshop and this discussion (as well as the views themselves) are reviewed here.

Boretsky argues that the marginal productivity of new capital investment in the United States declined during the 1960's, a decline which he attributes to a reduction in innovative activity. As evidence, he shows that new capital investment grew at a much more rapid rate than GNP during the years 1963-1969. There is, however, disagreement with the interpretation of this data. The problem is that new capital investment is a very cyclical data series, and the beginning and ending dates of the time period chosen by Boretsky correspond to the trough and peak years of the capital investment cycle. Boretsky's data, cyclically adjusted, do not appear to support the trend he believes is evident.[11]

Perhaps the most commented upon of Boretsky's data series has been his presentation of trade data showing a decline in the "favorable" trade balance of the United States (excess of exports over imports) in technology-intensive goods. The significance of these data rests in the concept of the product cycle, discussed in the previous chapter. It was argued that a nation in which new product innovation takes place will be likely to export the new products produced by the innovation early in the "life cycle" of these products but that as the products mature, the more likely it becomes that these products will be manufactured overseas. Thus, a nation that is continually innovating will also continually be exporting new or novel products. In principle, then, the innovative content of a nation's exports can serve as a measure of domestic innovative output.

The problem with using trade data as such a measure lies in the difficulty of determining what is or is not a new product. Using R&D expenditures as a percent of value added as a criterion, Boretsky classified manufacturing industries by two-digit SIC codes as being either "research and development intensive" or otherwise and then looked at the trade balances of all products in each of these two groupings. His findings were that the United States experienced a declining trade balance in the R&D intensive goods between the years. (See Table A-1.) This decline was advanced as evidence of a fall in the rate of U.S. innovative activity.

An objection to this approach is that the very broad aggregation of goods by industry as presented by Boretsky invariably mixes goods representing recent product innovations with those which do not. This level of aggregation, it was argued at the workshop, is so broad that it is impossible to determine whether or not the data really do support the assertion of a declining rate of new innovation. A partial solution to the aggregation problem is to examine trade data that are more finely disaggregated. Such data series, created by Regina Kelly, show less deterioration in the balance of U.S. trade in technology-intensive goods than do Boretsky's data.[12] (See Table A-2, which also presents the Boretsky data for purposes of comparison.) Kelly's data also show a recovery of the favorable balance of trade following the devaluation of the dollar in 1971.

Leaders of the scientific and engineering communities have also spoken out against the possible loss of technological leadership by the United States. Prime among these have been two of the most respected members of the community, Harvey Brooks and Jerome Wiesner.[13] The points of view of these two men are marked, however, by some notable differences. Wiesner emphasizes the ill effects of the reduction in federal funds allocated to R&D, noting that "research and development funds measured in real dollars have been shrinking (because they failed to keep up with costs) for several years...." Particularly hard hit, and particularly ominous for long-term U.S. interests

TABLE A-1 U.S. Merchandise Trade by Major Commodity Group, 1951-1972 ($ billions)

Commodity Group	1951-1955[a]	1962	1965	1971	1972
Agricultural products					
Exports	3.2	5.0	6.2	7.7	9.4
Imports	4.4	3.9	4.1	5.8	6.2
BALANCE	-1.2	1.1	2.1	1.9	3.2
Minerals, fuels, other raw materials					
Exports	1.3	2.1	2.6	3.8	4.3
Imports	3.3	4.5	5.4	7.9	10.1
BALANCE	-2.0	-2.4	-2.8	-4.1	-5.8
Nontechnology-intensive manufactured products					
Exports	3.7	3.5	4.4	6.3	7.1
Imports	1.9	5.1	7.4	14.6	17.8
BALANCE	1.8	-1.6	-3.0	-8.3	-10.7
Technology-intensive manufactured products					
Exports	6.6	10.2	13.0	24.2	26.6
Imports	0.9	2.5	3.9	15.9	19.9
BALANCE	5.7	7.7	9.1	8.3	6.7
All commodities					
Exports	15.5	21.7	27.5	44.1	49.8
Imports	10.9	16.5	21.4	45.6	55.6
BALANCE	4.6	5.2	6.1	-1.5	-5.8

[a]Average.

SOURCE: Compiled by Michael Boretsky from U.S. Department of Commerce data.

TABLE A-2 U.S. Trade Performance in Technology-Intensive Products, 1967-1975 ($ billions)

Definition	1967	1968	1969	1970	1971	1972	1973	1974	1975
Exports									
Boretsky Definition of "Technology-Intensive Product"	15.7	18.1	20.3	22.2	23.8	26.1	34.2	47.7	55.2
Kelly Definition of "Technology-Intensive Product"	8.2	9.7	10.8	12.4	13.3	13.7	18.5	25.9	29.2[a]
Imports									
Boretsky Definition of "Technology-Intensive Product"	6.8	9.2	11.1	12.7	15.6	19.5	23.7	28.2	27.3
Kelly Definition of "Technology-Intensive Product"	2.7	3.4	4.1	4.8	5.4	7.2	9.2	11.3	11.2[a]
Trade balance									
Boretsky Definition of "Technology-Intensive Product"	8.9	8.9	9.1	9.5	8.2	6.6	10.5	19.5	27.9
Kelly Definition of "Technology-Intensive Product"	5.5	6.3	6.7	7.6	7.9	6.5	9.3	14.6	18.0[a]

[a] Preliminary.

SOURCE: Compiled by Regina Kelly using U.S. Department of Commerce data.

in Wiesner's view is the serious financial plight of major research universities and other nonprofit research centers that (in Wiesner's words) "...perform most of the fundamental and exploratory research that is the foundation for technical innovation."

The position of Brooks is more moderate than that of Wiesner. Brooks emphasizes three forces behind the apparent demise of the leadership position of the United States in the development of new technology: (1) the technological inferiority of Europe in the years following World War II was "unnatural" and bound to disappear as the rebuilding of the European economy progressed; (2) the portion of the U.S. population that can be mobilized to undertake scientific and technological development has become fully utilized, so growth in the rate of technological innovation can no longer come about by means of mobilization of underutilized resources; and (3) the U.S. public has become somewhat concerned about the side effects of technology, and partly as a consequence of this concern, national priorities have shifted away from technological achievement and toward social welfare goals. Brooks believes that the latter two forces will soon affect the innovative output of other advanced nations. As is Wiesner, Brooks is concerned about the impact of reduced government financial support of R&D on the ability of the United States to generate new technology, but Brooks feels that the effect of these reductions per se will be minor in the long run in comparison to saturation of scientific manpower and changed national priorities.[14]

Neither Wiesner nor Brooks identify technology transfer from the United States as a major source of the technological problems of the United States. In Wiesner's words, "...I hope that we do not attempt to limit the flow of technological information from the United States (or any other country) if we do, in fact, discover the makings of a reverse technological gap."

In spite of the pessimism expressed by both Brooks and Wiesner over the present plight of U.S. technology, both men remain fundamentally optimistic about U.S. long-term technological capabilities. The opinion of both is that the United States has both the capability and the need to generate significant technological innovation in the future. Both, however, see the need (but to different degrees) for revised priorities if the potential of the United States is to be realized.

Not all commentators believe that the United States necessarily is suffering from a decline in ability to develop new technology. Yale economist Richard Cooper, for example, points out that loss of the American leadership position in the development of commercial technology has been a fear of U.S. policy makers since at least the early part of the twentieth century.[15] This fear, according to Cooper, abated during the 1950's and early 1960's, largely because the havoc and destruction of World War II temporarily eliminated several of the world's most innovative nations from the economic arena. The re-emergence of a fear of a declining U.S. position, in Cooper's view, stems partially from the re-emergence of Japan, Germany, and other nations as international economic powers and partially from the increasing rapidity with which technology can be transferred internationally. Cooper believes that the international economic system is entering an era in which half a dozen or so nations will share "pride of place" in the development of commercially applicable technology and that the speed with which any of these nations can imitate others' innovations will be so rapid that "any trade advantage accruing to the innovating country will be short-lived."

According to Cooper, none of this implies that the United States is losing its capacity to innovate. Two possible obstacles to U.S. international competitiveness are, however, identified by Cooper. The first is an unfavorable monetary environment (i.e., an overvalued U.S. dollar).[16] The second is a tendency for U.S. firms to become less cost conscious, in Cooper's view, than they have been historically. In this regard, Cooper shares a concern of J. H. Hollomon (see Chapter 2) and Harvey Brooks that the high concentration of U.S. engineers in the defense and aerospace industries has led to a propensity for U.S. technology to ignore issues of manufacturing efficiency in favor of developing products which are "overengineered," high-technology items. Cooper urges that U.S. firms must become more cost conscious.

According to Richard Nelson, a colleague of Richard Cooper, "the criterion of maintaining or achieving comprehensive technological leadership is neither a feasible nor a desirable criterion on which to base policy."[17] Nelson points out that the United States, in terms of technological capabilities, has long been "ahead on average" of other nations, but with the exception perhaps of the post-World War II years, the United States has never been the leader in every field. In many fields, the United States has been a follower rather than an innovator, and in Nelson's words, "it seems to have survived well."

In a paper prepared for the Woods Hole Workshop, E. M. Graham argued that there might be net benefits to the United States from a loss of its international role as the dominant source of technological innovation.[18] The argument was that the costs of being the innovator of new technology often are far in excess of the costs of imitating the innovation. If Richard Cooper is correct that half a dozen or so nations are beginning to share with the United States the role of being the source of the world's technological innovation and that the international diffusion of technology is proceeding more rapidly, the United States might increasingly find itself in the position of being able to import new technology more cheaply than it could be developed domestically. In the long run, argues Graham, this is a much more favorable position for the United States to be in than one which the United States unilaterally bears the cost of most innovation but shares the benefits internationally.

NOTES

1. See, for example, James Brian Quinn, "Technological Competition: Europe vs. U.S.," *Harvard Business Review*, July-Aug. 1966.
2. See C. P. Kindleberger, "An American Economic Climacteric?" *Challenge*, Jan.-Feb. 1974, pp. 35-44, and "Don't Look Back--They May Be Gaining on Us," in *Technological Innovation and Economic Development: Has the U.S. Lost the Initiative* (Washington, D.C.: Energy Research and Development Administration, 1976).
3. A Keynesian economist might, however, disagree strongly with this conclusion. If the marginal propensity of consumers to save is less than the marginal propensity of producers to invest, according to the standard Keynesian analysis, the producers will expand output until desired savings equal desired investment. The rate of investment might fall as expansion of output occurs, but aggregate investment will rise. A high marginal propensity for producers to invest might in turn be created by expectations that future demand for consumable goods will be high, that is, that marginal propensity to save is low. Thus, high rates of consumption, rather than high rates of saving, are the key to high aggregate investment.
4. See S. H. Hymer, *The International Operations of National Firms: A Study of Foreign Direct Investment* (Cambridge, Mass.: MIT Press, 1976); C. P. Kindleberger,

American Business Abroad (New Haven, Conn.: Yale University Press, 1969); Richard Caves, "International Corporations: The Industrial Economics of Foreign Investment," *Economics*, February 1971; Raymond Vernon, "The Location of Economic Activity," in J. H. Dunning, editor, *Economic Analysis and the Multinational Enterprise* (London: George Allen and Unwin, 1974) for propositions on the role of technology in the foreign direct investment activities of United States firms. For the British case, see John Seeley, *Expansion of England, 1883-1914*, Reprinted (New York: Macmillan Publishing Company, Inc., 1925), and A. K. Cairncross, *Home and Foreign Investment, 1870-1913* (New York: Cambridge University Press, 1929). See also D. C. M. Platt, *Finance, Trade, and Politics in British Foreign Policy 1815-1914* (Oxford, Engl.: Clarendon Press, 1968).

5. For the United States, see Mira Wilkins, *The Emergence of Multinational Enterprise* (Cambridge, Mass.: Harvard University Press, 1970) and "Multinational Enterprises: A Consideration of the Investment Strategies of Western Multinational Enterprises in the 19th and 20th Century, with Emphasis on the U.S. Corporation Abroad," Mimeo (Miami: Florida International University, 1970); for Germany, see Thorstein Veblen, *Imperial Germany and the Industrial Revolution* (1939; reprint ed., Ann Arbor: University of Michigan Press, 1966); Warner F. Brück, *Social and Economic History of Germany from Wilhelm II to Hitler, 1888-1938* (1939; reprint ed., New York: Russell and Russell, 1962); Gustav Stolper, *The German Economy, 1870 to the Present* (1940; English translation, New York: Harcourt, Brace, and World, 1967); and K. D. Barkin, *The Controversy Over German Industrialization, 1890-1902* (Chicago: University of Chicago Press, 1970).

6. See M. Y. Yoshino, *Japan's Multinational Enterprises* (Cambridge, Mass.: Harvard University Press, 1976) for one somewhat pessimistic appraisal of Japan's capabilities to innovate.

7. See L. F. Haber, *The Chemical Industry 1900-1930* (Oxford, Engl.: Clarendon Press, 1971), chaps. 2, 5.

8. One consequence of this is that both Japanese and German firms are increasingly extending their operations into the United States, often bringing with them their own best technology. See David S. McClain, "Foreign Investment in United States Manufacturing and the Theory of Direct Investment" (Ph.D. thesis, MIT, Cambridge, Mass., 1974); Edward M. Graham, "Oligopolistic Imitation and European Direct Investment in the United States" (D.B.A. thesis, Harvard University, Cambridge, Mass., 1974); and National

Research Council, *Technology Transfer from Foreign Direct Investment in the United States* (Washington, D.C.: National Academy of Sciences/National Academy of Engineering, 1976) for various views on this phenomenon.

9. See Robert Gilpin, *U.S. Power and the Multinational Corporation* (New York: Basic Books, 1974), chaps. 5, 6.

10. See Michael Boretsky, "Trends in U.S. Technology: A Political Economist's View," *American Scientist*, Jan.-Feb. 1975.

11. See "Comments on Dr. Boretsky's *American Scientist* Article" (internal memorandum, U.S. Department of Commerce, Washington, D.C., 1975) for comments on Boretsky's statistical methodology.

12. See Regina Kelly, "Alternative Measures of Technology-Intensive Trade," Mimeo (Washington, D.C.: U.S. Department of Commerce, August 1976).

13. See Harvey Brooks, "What's Happened to U.S. Lead in Technology?," *Harvard Business Review*, May-June 1972, and Jerome Wiesner, "Has the U.S. Lost Its Initiative in Technological Innovation?," *Technology Review*, July-August 1976. See also Nicholas Valéry, "The Declining Power of American Technology," *New Scientist*, July 1976.

14. Both Brooks and Wiesner acknowledge that most federal funding of R&D in the post-World War II era has been a result of space and defense programs. The two men disagree somewhat on the seriousness of the consequences of reductions of this funding. Wiesner believes that the Defense Department's funding of R&D historically has been highly effective, resulting in "spin-off" benefits to society that go far beyond narrow considerations of national defense. Brooks also believes that there are in fact important civilian spin-off benefits from space and defense programs, some of which are yet to come. Brooks also believes, however, that the space and defense programs entailed a large social "opportunity cost," because scarce technological resources were drawn away from the civilian sectors of the economy and into aerospace efforts.

15. See Richard N. Cooper, "Technology and U.S. Trade: A Historical Review," in *Technology and International Trade* (Washington, D.C.: National Academy of Engineering, 1971), pp. 3-17. Cooper quotes a passage written by Frank Taussig in 1915 which summarizes the contemporary views:

> The more machinery becomes automatic, the more readily it can be transplanted.... An American firm, it is said, will devise a new machine, and an export of the machine itself or of its products

will set in. Then some German will buy a specimen and reproduce the machine in his own country.... Soon not only will the exports cease, but the machine itself will be operated in Germany by low-paid labor, and the articles made by its aid will be sent back to the United States. Shoe machinery and knitting machinery have been cited in the illustration.

With the possible exception that today Germany would not be used as the example of the imitating nation, it is not infeasible that the above passage could have been written by a U.S. labor leader in 1976. The passage in fact appeared in F. W. Taussig, "Some Aspects of the Tariff Question," in *Selected Readings in International Trade and Tariff Problems* (Boston: Ginn and Company, 1921). (The views as expressed in the passage were not necessarily ones adhered to by Taussig himself.)

16. Cooper's concerns on this matter have abated somewhat following the devaluation of the dollar and the adoption of flexible exchange rates during 1971-1972.

17. See Richard R. Nelson, "'World Leadership,' the 'Technological Gap,' and National Science Policy," *Minerva*, July 1971, pp. 386-399.

18. See E. M. Graham, "Technological Innovation, the 'Technology Gap,' and U.S. Welfare: Some Observations" (unpublished background paper for the National Research Council/National Academy of Engineering Workshop on Technology and Trade, 1976). Available from the National Academy of Engineering, Office of the Foreign Secretary, Washington, D.C.

APPENDIX B   TRANSFER OF TECHNOLOGY TO DEVELOPING NATIONS: THE FINANCIAL ASPECTS

One workshop participant indicated that in his opinion a reason why the cost of technology has become a major issue for many developing countries is that these countries have experienced chronic balance-of-payments difficulties and increasing levels of international indebtedness. Balance-of-payments problems in these nations are caused by a number of factors, including imports in excess of exports and international debt-servicing requirements. (See Tables B-1 and B-2.) Problems of these nations recently have included increased prices for petroleum and increased requirements for imports of capital equipment.

The increasing international indebtedness of the developing countries can be attributed to several factors, of which the following two are probably the most significant: the need to borrow foreign exchange to pay for current

TABLE B-1 Net Trade Balance of Developing Nations Other Than OPEC[a] (U.S. $ billions)

	1973	1974	1975	1976[b]
Exports	67.70	98.25	95.49	25.03
Imports	80.13	130.53	137.96	34.00
Net imports	12.43	32.28	42.47	8.97

[a] Organization of Petroleum Exporting Countries.
[b] First quarter of 1976.

SOURCE: International Monetary Fund, *International Financial Statistics*, September 1976.

TABLE B-2 Net Trade Balance for Selected "Rapidly Industrializing" Developing Nations, 1975 (U.S. $ billions)

Nation	Exports	Imports	Net Imports
Brazil	8.66	13.56	4.90
Mexico	2.91	6.58	3.47
India	4.23	6.14	1.91
South Korea	5.08	7.27	2.19

SOURCE: International Monetary Fund, *International Financial Statistics*, September 1976.

account deficits and the need to finance public investment. The latter need stems from the fact that gross, domestically generated savings in most developing nations on the aggregate has been less than gross investment. (See Tables B-3 and B-4.)

There was concern expressed within the workshop that levels of indebtedness, and in particular private

TABLE B-3 External Outstanding Public Debt of Developing Nations, 1969, 1971, and 1973, at Year-End, and 1974 Estimate (U.S. $ billions)

	1969	1971	1973	1974[a]
Bilateral official	33.90	43.45	56.34	NA
Multilateral	11.42	16.43	24.08	NA
Private				
Suppliers	9.59	11.92	12.77	NA
Banks	4.60	8.06	17.83	28.7
Other	4.63	6.27	7.88	NA
TOTAL	64.14	86.13	118.89	151.4

NA = not applicable.
[a]Estimated.

SOURCE: International Bank for Reconstruction and Development, *Annual Report*, 1975.

TABLE B-4 Gross Investment and Gross Savings as a Percent of GNP

	1966-1970	1971	1972	1973
All developing countries[a]				
Gross investment	19.4	20.5	20.8	21.1
Gross national savings	17.0	18.1	19.1	20.8
Africa				
Gross investment	18.5	21.0	19.9	21.0
Gross national savings	15.3	15.9	16.7	20.8
East Asia				
Gross investment	19.6	21.8	21.3	22.0
Gross national savings	14.8	17.9	18.1	20.3
Middle East				
Gross investment	20.8	20.8	21.0	20.7
Gross national savings	21.2	24.6	26.0	33.4
South Asia				
Gross investment	14.4	15.8	16.2	15.9
Gross national savings	11.8	14.0	15.1	14.1
Western Hemisphere				
Gross investment	19.4	20.6	21.3	21.6
Gross national savings	17.8	17.7	18.6	19.7

[a]Eighty-six developing nations having loans with World Bank.

SOURCE: International Bank for Reconstruction and Development, *Annual Report*, 1975.

indebtedness, of developing countries were excessive. There was also fear expressed that some developing nations might not be able to meet debt-servicing requirements and thus default on the outstanding debt.

Despite these fears, it was felt that the international financial system is well equipped to handle any crisis that might arise from individual nations being unable in the short run to meet debt-servicing requirements. For example, should a nation be in danger of defaulting on private debt, it was felt that the International Monetary Fund (IMF) would surely come to the aid of that nation and prevent such a default from occurring. By extending credit to such a nation on favorable terms, the IMF could forestall a crisis.

In the long run, of course, such action would only

postpone the inevitable, were nations to become internationally indebted in excess of their long-term ability to service their debt. For this reason, it was believed that the IMF might, in certain cases, have to examine more closely the total international indebtedness of large borrowers to determine if prudent limits had been exceeded. Such an examination should encompass all levels of international debt, public and private. In cases where it was determined by the IMF that international indebtedness was excessive, the IMF could exercise any number of options open to it to force the offending nation to adopt more prudent policies.

Two further points were emphasized. First, the problems of most developing nations that experience debt-servicing problems are short term in nature; for example, when the price of a major exported commodity goes into a cyclical downturn, causing export earnings to decline. Such problems do not warrant special scrutiny by the IMF as described in the previous paragraph. Second, private bank loans to most developing nations in most cases are granted only after very careful analysis by the lending institution. Implicit in such an analysis, however, is the expectation that the IMF will act as a lender of last resort, so that the risk of naked default is considered to be quite low. The IMF should endeavor to determine if this expectation unduly lowers the risk averseness of the private lending institution so as to enable nations to overborrow from private sources. While there is no overt evidence to suggest that this is the case, it was felt that such a determination by the IMF should still be made.

SELECTED BIBLIOGRAPHY

American Economic Association. "Technical Progress, Capital Formation, and Economic Growth." *Papers*, May 1962.

Arrow, Kenneth J. "Comment." In *The Technology Factor in International Trade*, edited by R. Vernon. New York: National Bureau of Economic Research, 1970.

Baranson, Jack. "International Transfers of Industrial Technology by U.S. Firms and Their Implications for the U.S. Economy." In *Discussion Papers on International Trade, Foreign Investment, Employment*. Washington, D.C.: U.S. Department of Labor, Bureau of International Labor Affairs, Office of Foreign Economic Research, December 1976.

Barkin, K. D. *The Controversy Over German Industrialization, 1890-1902*. Chicago, Ill.: University of Chicago Press, 1970.

Behrman, Jack N. "The Multinational Enterprise and Economic Internationalism." *World Development* 3 (1975):845-856.

Bergendorff, Hans, and Strangent, Per. "Projections of Soviet Economic Growth and Defense Spending." Pp. 394-430 in *Soviet Economy in a New Perspective*, a compendium of papers submitted to the Joint Economic Committee, Congress of the United States, 94th Congress, 2d Session, 1976.

Bergsten, C. Fred. "Let's Avoid A Trade War." *Foreign Policy*, no. 23 (Summer 1976), pp. 24-31.

Berliner, Joseph S. "Prospects for Technological Progress." Pp. 431-446 in *Soviet Economy in a New Perspective*, a compendium of papers submitted to the Joint Economic Committee, Congress of the United States, 94th Congress, 2d Session, 1976.

Bhardan, P. K. "On Factor-Biased Technical Progress and International Trade." *The Journal of Political Economy*, August 1965.

Bitras, G., Lee, K., and Machlup, F. "Effects of Innovations on the Demand for and Earnings of Productive Factors." Mimeographed. Washington, D.C.: National Science Foundation, 1976.

Boretsky, Michael. "Trends in U.S. Technology: A Political Economist's View." *American Scientist*, January-February 1975.

"Brazil: The Aircraft Industry Irks U.S. Competitors." *Business Week*, October 11, 1976.

Brooks, Harvey. "What's Happened to U.S. Lead in Technology?" *Harvard Business Review*, May-June 1972.

Brougher, Jack. "U.S.S.R. Foreign Trade: A Greater Role for Trade with the West." Pp. 677-694 in *Soviet Economy in a New Perspective*, a compendium of papers submitted to the Joint Economic Committee, Congress of the United States, 94th Congress, 2d Session, 1976.

Brück, Warner F. *Social and Economic History of Germany from Wilhelm II to Hitler, 1888-1938*. 1939. Reprint. New York: Russell and Russell, 1962.

Bylinsky, Gene. *The Innovation Millionaires*. New York: Charles Scribner and Sons, 1976.

Cairncross, A. K. *Home and Foreign Investment, 1870-1913*. New York: Cambridge University Press, 1929.

Calmfors, Lars, and Rylander, Jan. "Economic Restrictions on Soviet Defense Expenditure." Pp. 377-393 in *Soviet Economy in a New Perspective*, a compendium of papers submitted to the Joint Economic Committee, Congress of the United States, 94th Congress, 2d Session, 1976.

Carey, D. N. "Soviet Agriculture: Recent Performance and Future Plans." Pp. 575-599 in *Soviet Economy in a New Perspective*, a compendium of papers submitted to the Joint Economic Committee, Congress of the United States, 94th Congress, 2d Session, 1976.

Caves, Richard. "International Corporations: The Industrial Economics of Foreign Investment." *Economica*, February 1971.

Caves, R., and Jones, R. *World Trade and Payments*. Boston, Mass.: Little, Brown, and Company, 1973.

Chipman, John S. "Induced Technical Change and Patterns of International Trade." In *The Technology Factor in International Trade*, edited by R. Vernon. New York: National Bureau of Economic Research, 1970.

Cohn, Stanley H. "Deficiencies in Soviet Investment Policies and the Technological Imperative." Pp. 447-459 in *Soviet Economy in a New Perspective*, a compendium of papers submitted to the Joint Economic Committee, Congress of the United States, 94th Congress, 2d Session, 1976.

Cooper, Richard N. "Technology and U.S. Trade: A Historical Review." Pp. 3-17 in *Technology and International Trade*. Washington, D.C.: National Academy of Engineering, 1971.

Corden, William M. "Economic Expansion and International Trade: A Geometric Approach." *Oxford Economic Papers* 8 (1956).

Davidson, W. H. "Patterns of Factor Saving Innovation in the Industrialized World." *European Economic Review*, December 1976.

Denison, Edward F. *Accounting for United States Economic Growth*. Washington, D.C.: Brookings Institution, 1974.

Denison, Edward F. "Comment on 'The Explanation of Productivity Change,' by D. Jorgenson and Z. Griliches." *Survey of Current Business*, May 1969.

Denison, Edward F. *Why Growth Rates Differ*. Washington, D.C.: Brookings Institution, 1967.

Diebold, William, Jr. "The Role of the Western Multinationals in East-West Cooperation." Pp. 134-135 in *Perspektiven und Probleme Wirtschaftlicher Zusammenarbeit Zwischen Ost-und Westeuropa*. Deutsches Institut Für Wirtschaftsforschung, Sonderheft 114, 1976.

Domar, Evsey. "On the Measurement of Technological Change." *Economic Journal*, December 1961.

Eckaus, Richard. *Appropriate Technologies for Developing Countries*. Washington, D.C.: National Academy of Sciences, 1977.

Edwards, Charles C. "The Role of Government and F.D.A. Regulations in Drug R&D." *Research Management*, March 1974.

Enos, John. "Invention and Innovation in the Petroleum Refining Industry." In *The Rate and Direction of Inventive Activity*, edited by R. R. Nelson. Princeton, N.J.: Princeton University Press, 1962.

Ericson, Paul. "Soviet Efforts to Increase Exports of Manufactured Products to the West." Pp. 709-726 in *Soviet Economy in a New Perspective*, a compendium of papers submitted to the Joint Economic Committee, Congress of the United States, 94th Congress, 2d Session, 1976.

"Export Licensing of Advanced Technology: A Review." Hearings before the Subcommittee on International Trade and Commerce of the Committee on International Relations, House of Representatives, 94th Congress, 2d Session, March 1976.

Farrell, John, and Ericson, Paul. "Soviet Trade and Payments with the West." Pp. 727-738 in *Soviet Economy in a New Perspective*, a compendium of papers submitted to the Joint Economic Committee, Congress of the United States, 94th Congress, 2d Session, 1976.

Findlay, R. and Grubert, H. "Factor Intensities, Technological Progress, and the Terms of Trade." Pp. 111-121 in *Oxford Economic Papers*, 1959. Reprinted in *International Trade: Selected Readings*, edited by J. Bhagwati. New York: Penguin Books, Inc., 1967.

Flender, J. O., and Morse, Richard. *The Role of New Technical Enterprises in the U.S. Economy*. Cambridge, Mass.: MIT Development Foundation, 1975.

Franko, Larry. *The European Multinationals*. New York: Harper and Row, 1976.

Gillette, Dean. "Innovation Under Regulation." Paper presented to Panel on The Effect of Government Antitrust Action and Regulation on Technological Innovation: The Issues, at Annual Meeting of the American Association for the Advancement of Science, Washington, D.C., February 20, 1976. Mimeographed.

Gilpin, Robert. *Technology, Economic Growth, and International Competitiveness*. A report used for the Subcommittee on Economic Growth of the Joint Economic Committee of the U.S. Congress. Washington, D.C.: U.S. Government Printing Office, July 9, 1975.

Gilpin, Robert. *U.S. Power and the Multinational Corporation*. New York: Basic Books, 1974.

Goldfinger, Nathan. "A Labor View of Foreign Investment and Trade Issues." In *International Trade and Finance*, edited by R. E. Baldwin and J. D. Richardson. Boston, Mass.: Little, Brown and Company, 1974.

Goldman, M. I. "Autarchy or Integration: The U.S.S.R. and The World Economy." Pp. 81-96 in *Soviet Economy in a New Perspective*, a compendium of papers submitted to the Joint Economic Committee, Congress of the United States, 94th Congress, 2d Session, 1976.

Goldman, M. I. *Detente and Dollars: Doing Business with the Soviet Union*. New York: Basic Books, 1975.

Graham, Edward M. "Oligopolistic Imitation and European Direct Investment in the United States." D.B.A. thesis, Cambridge, Mass., Harvard University, 1974.

Graham, Edward M. "Technological Innovation, the 'Technology Gap,' and U.S. Welfare: Some Observations." Unpublished background paper for the National Research Council/National Academy of Engineering Workshop on Technology and Trade, 1976. (Available from Office of the Foreign Secretary, National Academy of Engineering, Washington, D.C.)

Greenfield, Stanley M. "Incentives and Disincentives of of EPA Regulations." *Research Management,* March 1974.

Griliches, Zvi. "Comment." *American Economic Review Papers and Proceedings*, May 1962.

Gruber, W., Mehta, D., and Vernon, R. "The R&D Factor in International Trade and International Investment of United States Industries." *The Journal of Political Economy*, February 1967.

Gustafson, W. Eric. "R&D, New Products, and Productivity Change." *American Economic Review Papers and Proceedings*, May 1962.

Haber, L. F. *The Chemical Industry 1900-1930*. Oxford, Engl.: Clarendon Press, 1971.

Hager, E. E. *The Economics of Development*. Homewood, Ill.: Richard D. Irwin, 1968.

Hanson, Philip. "International Technology Transfer From the West to the U.S.S.R." Pp. 786-812 in *Soviet Economy in a New Perspective*, a compendium of papers submitted to the Joint Economic Committee, Congress of the United States, 94th Congress, 2d Session, 1976.

Hardt, John P. "Summary." Pp. ix-xxxix in *Soviet Economy in a New Perspective*, a compendium of papers submitted to the Joint Economic Committee, Congress of the United States, 94th Congress, 2d Session, 1976.

Heston-Sands, Mary, and Hope, Lawrence L. "Strategy and Planning in a Turbulent Environment: The Ethical Pharmaceutical Industry." S.M. thesis, MIT Sloan School of Management, Cambridge, Mass., 1976.

Hicks, J. R. "An Inaugural Lecture." *Oxford Economic Papers* 5 (1953).

Hicks, J. R. *The Theory of Wages*. New York: Macmillan Publishing Company, Inc., 1935.

Hollomon, J. Herbert. "America's Technological Dilemma." *Technology Review*, July-August 1971.

Holzman, Franklyn D. *Foreign Trade Under Central Planning*. Cambridge, Mass.: Harvard University Press, 1974.

Horst, Thomas. "American Taxation of Multinational Corporations." Mimeographed. Medford, Mass.: Fletcher School of Law and Diplomacy, Tufts University, September 1975.

Hymer, S. H. *The International Operations of National Firms: A Study of Foreign Direct Investment*. Cambridge, Mass.: MIT Press, 1976.

"I.B.M.'s $5,000,000,000 Gamble." *Fortune*, September 1966.

International Bank for Reconstruction and Development. *Annual Report, 1975*. Washington, D.C.: International Bank for Reconstruction and Development, 1975.

International Bank for Reconstruction and Development. *Trends in Developing Countries*. Washington, D.C.: International Bank for Reconstruction and Development, 1973.

International Monetary Fund. *International Financial Statistics*. Washington, D.C.: International Monetary Fund, September 1976.

Jecquier, N., ed. *Appropriate Technology: Problems and Promises*. Paris, France: Organization for Economic Cooperation and Development, 1976.

Jewkes, J., Sawers, D., and Stillerman, R. *The Sources of Invention*. New York: Macmillan Publishing Company, Inc., 1969.

Johnson, Harry G. *Economic Expansion and International Trade*. Manchester, England: Manchester School of Economics and Social Studies, 1955.

Johnson, Harry G. "Technological Change and Comparative Advantage: An Advanced Country's Viewpoint." *Journal of World Trade Law* 9 (January-February 1975).

Johnson, Harry G. "The Efficiency and Welfare Implications of the International Corporation." In *The International Corporation*, edited by C. P. Kindleberger. Cambridge, Mass.: MIT Press, 1970.

Jorgenson, Dale, and Griliches, Zvi. "The Explanation of Productivity Change." *The Review of Economic Studies*, July 1967.

Karchere, A. J. "The Effect of Transnational Companies on the U.S. Economy, and Future Prospects." Pp. 69-81 in *The International Essays for Business Decision Makers*, edited by M. B. Winchester. Proceedings, Annual International Trade Conference of the Southwest, May 18, 1976, Dallas, Texas. Dallas, Texas: School of Business Administration, Southern Methodist University, 1976.

Katz, S. Stanley. "The Developing World and U.S. Trade." In *The International Essays for Business Decision Makers*, edited by M. B. Winchester. Proceedings, Annual International Trade Conference of the Southwest, May 18, 1976, Dallas, Texas. Dallas, Texas: School of Business Administration, Southern Methodist University, 1976.

Keesing, D. B. "The Impact of Research and Development on United States Trade." *The Journal of Political Economy*, February 1967.

Kelly, Regina. "Alternative Measures of Technology-Intensive Trade." Mimeographed. Washington, D.C.: U.S. Department of Commerce, August 1976.

Kemp, Murray C. *The Pure Theory of International Trade and Investment*. New York: Prentice Hall, 1969.

Kendrick, John W. *Postwar Productivity Trends in the United States*. New York: National Bureau of Economic Research, 1973.

Kendrick, John W. "Productivity Trends and Prospects." Paper prepared for the Joint Economic Committee, Congress of the United States, 94th Congress, 2d Session, June 1976. Mimeographed.

Kennedy, C. "Induced Bias in Innovation and the Theory of Distribution." *Economic Journal*, September 1964.

Kindleberger, C. P. *American Business Abroad*. New Haven, Conn.: Yale University Press, 1969.

Kindleberger, C. P. "An American Economic Climacteric?" *Challenge*, January-February 1974, pp. 35-44.

Kindleberger, C. P. "Don't Look Back--They May Be Gaining On Us." In *Technological Innovation and Economic Development: Has the U.S. Lost the Initiative?* Washington, D.C.: Energy Research and Development Administration, 1976.

Kindleberger, C. P. *The World in Depression, 1929-1939*. Berkeley and Los Angeles, Calif.: University of California Press, 1970.

Kohn, Martin J. "Developments in Soviet-Eastern European Terms of Trade, 1971-75." In *Soviet Economy in a New Perspective*, a compendium of papers submitted to the Joint Economic Committee, Congress of the United States, 94th Congress, 2d Session, 1976.

Kreinin, Mordechai. "The Leontief Scarce-Factor Paradox." *American Economic Review*, March 1965.

Kuznets, Simon. "Inventive Activity's Problems of Definition and Measurement." In *The Rate and Direction of Inventive Activity*, edited by R. R. Nelson. Princeton, N.J.: Princeton University Press, 1962.

Landau, Ralph. Statement on "Financial and Capacity Needs." Hearings before the Joint Economic Committee, Congress of the United States, 93rd Congress, 2d Session, October 2, 1974, pp. 112-133.

Landau, Ralph, and Mendolia, Arthur I. "An American View of Chemical Investment Patterns in the Era of High Energy Costs." *Chemistry and Industry*, December 6, 1975.

"Lestoil: The Road Back." *Business Week*, June 15, 1963.

Levine, Herbert S. "Economic Interdependence and the U.S.-Soviet Relationship." Paper presented at the Third National Security Affairs Conference, National Defense University, Washington, D.C., July 12-14, 1976. Mimeographed.

Long, T. Dixon. "Japan's Technological Policy: Challenge or Warning?" Paper presented at the conference on

"Technological Innovation and Economic Development: Has the U.S. Lost the Initiative?" Washington, D.C., April 20, 1976. Mimeographed.

Maclaurin, W. R. *Invention and Innovation in the Radio Industry*. New York: Macmillan Publishing Company, Inc., 1949.

Mansfield, Edwin. *Technological Change--An Introduction to a Vital Area of Modern Economics*. New York: Norton Publishing Company, 1971.

Mansfield, E., Rapaport, J., Romeo, A., Wagner, S., and Beardsley, G. "Social and Private Rates of Return from Industrial Innovations." *Quarterly Journal of Economics* 2 (May 1977).

Massachusetts Institute of Technology. *A Factbook Concerning the Relationship Between Technology and Trade*. Cambridge, Mass.: Center for Policy Alternatives, Massachusetts Institute of Technology, 1976. (Will be available from National Technical Information Service, Springfield, Virginia 22161.)

Maynes, Charles William. "A U.N. Policy for the Next Administration." *Foreign Affairs* 54 (July 1976).

McClain, David S. "Foreign Investment in United States Manufacturing and the Theory of Direct Investment." Ph.D. thesis, Massachusetts Institute of Technology, Cambridge, Mass., 1974.

Mill, J. S. *Principles of Political Economy*. New York: D. Appleton Publishing Company, 1890.

Morley, S. A., and Smith, G. W. "Limited Search and the Technology Choices of Multinational Firms in Brazil." *Quarterly Journal of Economics* 91, no. 2 (May 1977).

Morse, Richard S. "Innovative Technology: What Is Its Impact on U.S. Economy?" *Professional Engineer*, August 1976.

Mueller, W. F. "The Origins of the Basic Inventions Underlying DuPont's Major Product and Process Innovations, 1920 to 1950." In *The Rate and Direction of Inventive Activity*, edited by R. R. Nelson. Princeton, N.J.: Princeton University Press, 1962.

Musgrave, Peggy B. *Direct Investment Abroad and the Multinationals: Effects on the United States Economy*. U.S. Senate Subcommittee on Foreign Relations. Washington, D.C.: U.S. Government Printing Office, August 1975.

Musgrave, Peggy B. "Tax Preferences to Foreign Investment." In *Economics of Federal Subsidy Programs, Part 2-International Subsidies*. Joint Economic Committee, 92d Congress, 2d Session 176, Washington, D.C., 1972.

Nadiri, M. I. "Some Approaches to the Theory and Measurement of Total Factor Productivity: A Survey." *Journal of Economic Literature*, December 1970.

Nasbeth, L., and Ray, G. F. *The Diffusion of New Industrial Processes*. New York: Cambridge University Press, 1974.

National Academy of Engineering. *Meeting the Challenge of Industrialization*. Washington, D.C.: National Academy of Sciences, 1973. (Available from National Technical Information Service, Springfield, Virginia 22161, NTIS #PB-228-348, $5.75.)

National Academy of Engineering. *U.S. Technology and International Trade*. Proceedings of the Technical Session at the Eleventh Annual Meeting, April 23-24, 1975. Washington, D.C.: National Academy of Sciences, 1976.

National Research Council. *Technology Transfer From Foreign Direct Investment in the United States*. Washington, D.C.: National Academy of Sciences, 1976.

National Science Board. *Science Indicators 1974*. Washington, D.C.: National Science Foundation, 1974.

National Science Foundation. *National Patterns of R and D Resources: Funds and Manpower in the United States*. Washington, D.C.: National Science Foundation, 1975.

National Science Foundation. *Technological Innovation and Federal Government Policy: Research and Analysis of the Office of National R&D Assessment*. Washington, D.C.: National Science Foundation, 1976.

Nau, Henry. *Technology Transfer and U.S. Foreign Policy*. New York: Praeger Publishers, 1976.

Nelson, Richard R. "'World Leadership,' the 'Technological Gap,' and National Science Policy." *Minerva*, July 1971, pp. 386-399.

Neuman, George R. "The Direct Labor Market Effects of the Trade Adjustment Assistance Program: The Evidence from the T.A.A. Survey." Unpublished paper for the U.S. Department of Labor, Washington, D.C., 1977.

Neuman, G. R., "An Evaluation of the Trade Adjustment Assistance Program." Report to the U.S. Department of Labor, Washington, D.C., 1976. Mimeographed.

Organization for Economic Cooperation and Development. *Expenditure Trends in OECD Countries, 1960-1980*. Paris, France: Organization for Economic Cooperation and Development, 1972.

Organization for Economic Cooperation and Development. "Report by OECD's Manpower and Social Affairs Committee." *OECD Observer*, March-April 1976.

Organization for Economic Cooperation and Development. *The Growth of Output, 1960-1980*. Paris, France:

Organization for Economic Cooperation and Development, 1970.

Pack, H. "The Substitution of Labour for Capital in Kenyan Manufacturing." *Economic Journal*, March 1976.

Peck, M. J. "Inventions in the Postwar American Aluminum Industry." In *The Rate and Direction of Inventive Activity*, edited by R. R. Nelson. Princeton, N.J.: Princeton University Press, 1962.

Piore, M. J. "The Impact of the Labor Market Upon the Design and Selection of Productive Techniques Within the Manufacturing Plant." *Quarterly Journal of Economics*, November 1968.

Platt, D. C. M. *Finance, Trade, and Politics in British Foreign Policy, 1815-1914*. Oxford, England: Clarendon Press, 1968.

Quinn, James Brian. "Technological Competition: Europe vs. U.S." *Harvard Business Review*, July-August 1966.

Robinson, J. "The Classification of Inventions." *Review of Economic Studies* 5 (1937-1938).

Rosenberg, Nathan. "Factors Affecting the Payoff to Technological Innovation." Mimeographed. Washington, D.C.: National Science Foundation, 1976.

Sadusk, Joseph F., Jr. "The Effect of Drug Regulation on the Development of New Drugs." In *Principles and Techniques of Human Research and Therapeutics*, volume 1, edited by F. Gilbert McMahon. Mount Kisco, N.Y.: Futura Publishing Company, 1974.

Salter, W. E. G. *Productivity and Technical Change*. New York: Cambridge University Press, 1964.

Sarett, Lewis A. "FDA Regulations and their Influence on Future R&D." *Research Management*, March 1974.

Scherer, F. M. "Firm Size and Patented Inventions." *American Economic Review*, December 1965.

Scherer, F. M. *Industrial Market Structure and Economic Performance*. Chicago, Ill.: Rand McNally and Company, 1970.

Scherer, F. M. "The Development of the TD-X and TD-2 Microwave Radio Relay Systems in Bell Telephone Laboratories." Mimeographed. Harvard Business School Case Study, 1960. (Available from Intercollegiate Case Clearinghouse at Harvard Business School, Cambridge, Mass.)

Schmookler, Jacob. *Invention and Economic Growth*. Cambridge, Mass.: Harvard University Press, 1966.

Schumpeter, Joseph A. *Business Cycles*. 2 volumes. New York and London: McGraw-Hill, 1939.

Schumpeter, Joseph A. *History of Economic Analysis*. New York: Oxford University Press, 1954.

Schumpeter, Joseph A. "The Analysis of Economic Change." In *Readings in Business Cycle Theory*. London, England: Blakiston, 1944.

Seeley, John. *Expansion of England, 1883-1914*. Reprint. New York: Macmillan Publishing Company, Inc., 1925.

Servan-Schreiber, J. J. *The American Challenge*. Translated from the French, *Le Défi Américain*, 1967. New York: Atheneum Publishers, 1968.

Smith, Bruce L. R., and Karlesky, Joseph J. *The State of Academic Science, The Universities in the Nation's Research Effort*. New Rochelle, N.Y.: Change Magazine Press, 1977.

Solomon, R. F., and Forsyth, D. J. C. "Substitution of Labour for Capital in the Foreign Sector: Some Further Evidence." *Economic Journal*, June 1977.

Solow, R. M. *Capital Theory and the Rate of Return*. London, England: North-Holland Publishing Co., Ltd., 1964.

Solow, R. M. "Investment and Technical Progress." In *Mathematical Methods in the Social Sciences*, edited by K. J. Arrow, S. Karlin, and P. Suppes. Stanford, Calif.: Stanford University Press, 1969.

Steele, Lowell W. *Innovation in Big Business*. New York: American Elsevier Publishing Company, 1975.

Stobaugh, Robert B. "The Product Life Cycle, U.S. Exports, and International Investment." Ph.D. thesis, Harvard University, Cambridge, Mass., 1968.

Stobaugh, Robert B., ed. *Nine Investments Abroad and their Impact at Home*. Cambridge, Mass.: Harvard University Press, 1976.

Stolper, Gustav. *The German Economy, 1870 to the Present*. Originally published, 1940. English translation. New York: Harcourt, Brace and World, 1967.

Sturmey, S. G. *The Economic Development of Radio*. London, England: Duckworth, 1958.

"Survey of Governmental Regulation." *Business Week*, April 4, 1977, pp. 42, 43+.

Taussig, F. W. "Some Aspects of the Tariff Question." In *Selected Readings in International Trade and Tariff Problems*, edited by F. W. Taussig. Boston, Mass.: Ginn and Company, 1921.

Thurow, L. C., and White, Halbert. "Optimum Trade Restrictions and their Consequences." *Econometrica*, July 1976.

Tsurumi, Yoshi. *The Japanese are Coming*. Cambridge, Mass.: Ballinger Publishing Company, 1976.

U.S. Department of Defense. *An Analysis of Export Control of U.S. Technology--A DOD Perspective: A Report of the*

Defense Science Board Task Force on Export of Technology. Washington, D.C.: Office of the Director of Defense Research and Engineering, U.S. Department of Defense, 1976.

U.S. Energy Research and Development Administration. *The United States in the Changing World Economy.* Washington, D.C.: U.S. Energy Research and Development Administration, 1971.

U.S. Tariff Commission. *Report to the Committee on Finance of the United States Senate on Implications of Multinational Firms for World Trade and Investment and for U.S. Trade and Labor.* Washington, D.C.: U.S. Tariff Commission, 1973.

Valéry, Nicholas. "The Declining Power of American Technology." *New Scientist*, July 1976.

Veblen, Thorstein. *Imperial Germany and the Industrial Revolution.* 1939. Reprint. Ann Arbor, Michigan: University of Michigan Press, 1966.

Vernon, Raymond. "International Investment and International Trade in the Product Cycle." *Quarterly Journal of Economics*, May 1966.

Vernon, Raymond. *Sovereignty at Bay.* New York: Basic Books, 1970.

Vernon, Raymond. "The Location of Economic Activity." In *Economic Analysis and the Multinational Enterprise*, edited by J. H. Dunning. London, England: George Allen and Unwin, 1974.

Vernon, Raymond, and Goldman, M. I. "U.S. Policies in the Sale of Technology to the U.S.S.R." Mimeographed. Cambridge, Mass.: Center for International Affairs, Harvard University, 1974.

Wardell, William M., and Lasagna, Louis. *Regulation and Drug Development.* Washington, D.C.: American Enterprise Institute for Public Policy Research, 1975.

Wells, Louis T. "Economic Man and Engineering Man: Choice of Technology in a Low Wage Country." *Public Policy*, Spring 1973.

Wells, Louis T. "Test of the Product Cycle Model of International Trade." *Quarterly Journal of Economics*, February 1969.

Wells, Louis T., ed. *The Product Life Cycle and International Trade.* Cambridge, Mass.: Harvard University Press, 1973.

"Where Private Industry Puts Its Research Money." *Business Week*, June 26, 1976.

Wiesner, Jerome. "Has the U.S. Lost Its Initiative in Technological Innovation?" *Technology Review*, July-August 1976.

Wilkins, Mira. "Multinational Enterprises: A Consideration of the Investment Strategies of Western Multinational Enterprises in the 19th and 20th Century, with Emphasis on the U.S. Corporation Abroad." Mimeographed. Miami, Florida: Florida International University, 1970.

Wilkins, Mira. *The Emergence of Multinational Enterprise*. Cambridge, Mass.: Harvard University Press, 1970.

Wilkins, Mira. *The Maturing of Multinational Enterprise*. Cambridge, Mass.: Harvard University Press, 1974.

Winpisinger, William W. "Remarks at the U.S. Department of State National Meeting on Science and Technology," Washington, D.C., November 17, 1976. Mimeographed. (Available from International Association of Machinists and Aerospace Workers, Washington, D.C.)

Yoshino, Michal Y. *Japan's Multinational Enterprises*. Cambridge, Mass.: Harvard University Press, 1976.